SpeedPro Series

THE COMPETITION CAR

DATA
LOGGING
MANUAL

Other great books from Veloce –

Speedpro Series
4-cylinder Engine – How To Blueprint & Build A Short Block For High Performance (Hammill)
Alfa Romeo DOHC High-performance Manual (Kartalamakis)
Alfa Romeo V6 Engine High-performance Manual (Kartalamakis)
BMC 998cc A-series Engine – How To Power Tune (Hammill)
1275cc A-series High-performance Manual (Hammill)
Camshafts – How To Choose & Time Them For Maximum Power (Hammill)
Competition Car Datalogging Manual, The (Templeman)
Cylinder Heads – How To Build, Modify & Power Tune Updated & Revised Edition (Burgess & Gollan)
Distributor-type Ignition Systems – How To Build & Power Tune (Hammill)
Fast Road Car – How To Plan And Build Revised & Updated Colour New Edition (Stapleton)
Ford SOHC 'Pinto' & Sierra Cosworth DOHC Engines – How To Power Tune Updated & Enlarged Edition (Hammill)
Ford V8 – How To Power Tune Small Block Engines (Hammill)
Harley-Davidson Evolution Engines – How To Build & Power Tune (Hammill)
Holley Carburetors – How To Build & Power Tune Revised & Updated Edition (Hammill)
Jaguar XK Engines – How To Power Tune Revised & Updated Colour Edition (Hammill)
MG Midget & Austin-Healey Sprite – How To Power Tune Updated & Revised Edition (Stapleton)
MGB 4-cylinder Engine – How To Power Tune (Burgess)
MGB V8 Power – How To Give Your, Third Colour Edition (Williams)
MGB, MGC & MGB V8 – How To Improve (Williams)
Mini Engines – How To Power Tune On A Small Budget Colour Edition (Hammill)
Motorcycle-engined Racing Car – How To Build (Pashley)
Motorsport – Getting Started in (Collins)
Nitrous Oxide High-performance Manual, The (Langfield)
Rover V8 Engines – How To Power Tune (Hammill)
Sportscar/kitcar Suspension & Brakes – How To Build & Modify Enlarged & Updated 2nd Edition (Hammill)
SU Carburettor High-performance Manual (Hammill)
Suzuki 4x4 – How To Modify For Serious Off-road Action (Richardson)
Tiger Avon Sportscar – How To Build Your Own Updated & Revised 2nd Edition (Dudley)
TR2, 3 & TR4 – How To Improve (Williams)
TR5, 250 & TR6 – How To Improve (Williams)
TR7 & TR8 – How To Improve (Williams)
V8 Engine – How To Build A Short Block For High Performance (Hammill)
Volkswagen Beetle Suspension, Brakes & Chassis – How To Modify For High Performance (Hale)
Volkswagen Bus Suspension, Brakes & Chassis – How To Modify For High Performance (Hale)
Weber DCOE, & Dellorto DHLA Carburetors – How To Build & Power Tune 3rd Edition (Hammill)

Those Were The Days ... Series
Alpine Trials & Rallies 1910-1973 (Pfundner)
Austerity Motoring (Bobbitt)
Brighton National Speed Trials (Gardiner)
British Police Cars (Walker)
British Woodies (Peck)
Dune Buggy Phenomenon (Hale)
Dune Buggy Phenomenon Volume 2 (Hale)
Hot Rod & Stock Car Racing in Britain In The 1980s (Neil)
MG's Abingdon Factory (Moylan)
Motor Racing At Brands Hatch In The Seventies (Parker)
Motor Racing At Crystal Palace (Collins)
Motor Racing At Goodwood In The Sixties (Gardiner)
Motor Racing At Nassau In The 1950s & 1960s (O'Neil)
Motor Racing At Oulton Park In The 1960s (Mcfadyen)
Motor Racing At Oulton Park In The 1970s (Mcfadyen)
Three Wheelers (Bobbitt)

Enthusiast's Restoration Manual Series
Citroën 2CV, How To Restore (Porter)
Classic Car Bodywork, How To Restore (Thaddeus)
Classic Car Electrics (Thaddeus)
Classic Cars, How To Paint (Thaddeus)
Reliant Regal, How To Restore (Payne)
Triumph TR2/3/3A, How To Restore (Williams)
Triumph TR4/4A, How To Restore (Williams)
Triumph TR5/250 & 6, How To Restore (Williams)
Triumph TR7/8, How To Restore (Williams)
Volkswagen Beetle, How To Restore (Tyler)
VW Bay Window Bus (Paxton)
Yamaha FS1-E, How To Restore (Watts)

Essential Buyer's Guide Series
Alfa GT (Booker)
Alfa Romeo Spider Giulia (Booker & Talbott)
BMW GS (Henshaw)
BSA Bantam (Henshaw)
BSA Twins (Henshaw)
Citroën 2CV (Paxton)
Citroën ID & DS (Heilig)
Fiat 500 & 600 (Bobbitt)
Jaguar E-type 3.8 & 4.2-litre (Crespin)
Jaguar E-type V12 5.3-litre (Crespin)
Jaguar/Daimler XJ6, XJ12 & Sovereign (Crespin)
Jaguar XJ-S (Crespin)
MGB & MGB GT (Williams)
Mercedes-Benz 280SL-560DSL Roadsters (Bass)
Mercedes-Benz 'Pagoda' 230SL, 250SL & 280SL Roadsters & Coupés (Bass)
Morris Minor & 1000 (Newell)
Porsche 928 (Hemmings)
Rolls-Royce Silver Shadow & Bentley T-Series (Bobbitt)
Subaru Impreza (Hobbs)
Triumph Bonneville (Henshaw)
Triumph TR6 (Williams)
VW Beetle (Cservenka & Copping)
VW Bus (Cservenka & Copping)

Auto-Graphics Series
Fiat-based Abarths (Sparrow)
Jaguar MKI & II Saloons (Sparrow)
Lambretta Li Series Scooters (Sparrow)

Rally Giants Series
Audi Quattro (Robson)
Austin Healey 100-6 & 3000 (Robson)
Fiat 131 Abarth (Robson)
Ford Escort MkI (Robson)
Ford Escort RS Cosworth & World Rally Car (Robson)
Ford Escort RS1800 (Robson)
Lancia Stratos (Robson)
Peugeot 205 T16 (Robson)
Subaru Impreza (Robson)

General
1½-litre GP Racing 1961-1965 (Whitelock)
AC Two-litre Saloons & Buckland Sportscars (Archibald)
Alfa Romeo Giulia Coupé GT & GTA (Tipler)
Alfa Romeo Montreal – The Essential Companion (Taylor)
Alfa Tipo 33 (McDonough & Collins)
Alpine & Renault – The Development Of The Revolutionary Turbo F1 Car 1968 to 1979 (Smith)
Anatomy Of The Works Minis (Moylan)
Armstrong-Siddeley (Smith)
Autodrome (Collins & Ireland)
Automotive A-Z, Lane's Dictionary Of Automotive Terms (Lane)
Automotive Mascots (Kay & Springate)
Bahamas Speed Weeks, The (O'Neil)
Bentley Continental, Corniche And Azure (Bennett)
Bentley MkVI, Rolls-Royce Silver Wraith, Dawn & Cloud/Bentley R & S-Series (Nutland)
BMC Competitions Department Secrets (Turner, Chambers Browning)
BMW 5-Series (Cranswick)
BMW Z-Cars (Taylor)
Britains Farm Model Balers & Combines 1967 to 2007 (Pullen)
British 250cc Racing Motorcycles (Pereira)
British Cars, The Complete Catalogue Of, 1895-1975 (Culshaw & Horrobin)
BRM – A Mechanic's Tale (Salmon)
BRM V16 (Ludvigsen)
BSA Bantam Bible, The (Henshaw)
Bugatti Type 40 (Price)
Bugatti 46/50 Updated Edition (Price & Arbey)
Bugatti T44 & T49 (Price & Arbey)
Bugatti 57 2nd Edition (Price)
Caravans, The Illustrated History 1919-1959 (Jenkinson)
Caravans, The Illustrated History From 1960 (Jenkinson)
Carrera Panamericana, La (Tipler)
Chrysler 300 – America's Most Powerful Car 2nd Edition (Ackerson)
Chrysler PT Cruiser (Ackerson)
Citroën DS (Bobbitt)
Cliff Allison – From The Fells To Ferrari (Gauld)
Cobra – The Real Thing! (Legate)
Cortina – Ford's Bestseller (Robson)
Coventry Climax Racing Engines (Hammill)
Daimler SP250 New Edition (Long)
Datsun Fairlady Roadster To 280ZX – The Z-Car Story (Long)
Dino – The V6 Ferrari (Long)
Dodge Charger – Enduring Thunder (Ackerson)
Dodge Dynamite! (Grist)
Donington (Boddy)
Draw & Paint Cars – How To (Gardiner)
Drive On The Wild Side, A – 20 Extreme Driving Adventures From Around The World (Weaver)
Ducati 750 Bible, The (Falloon)
Ducati 860, 900 And Mille Bible, The (Falloon)
Dune Buggy, Building A – The Essential Manual (Shakespeare)
Dune Buggy Files (Hale)
Dune Buggy Handbook (Hale)
Edward Turner: The Man Behind The Motorcycles (Clew)
Fiat & Abarth 124 Spider & Coupé (Tipler)
Fiat & Abarth 500 & 600 2nd Edition (Bobbitt)
Fiats, Great Small (Ward)
Fine Art Of The Motorcycle Engine, The (Peirce)
Ford F100/F150 Pick-up 1948-1996 (Ackerson)
Ford F150 Pick-up 1997-2005 (Ackerson)
Ford GT – Then, And Now (Streather)
Ford GT40 (Legate)
Ford In Miniature (Olson)
Ford Model Y (Roberts)
Ford Thunderbird From 1954, The Book Of The (Long)
Forza Minardi! (Vigar)
Funky Mopeds (Skelton)
Gentleman Jack (Gauld)
GM In Miniature (Olson)
GT – The World's Best GT Cars 1953-73 (Dawson)
Hillclimbing & Sprinting – The Essential Manual (Short & Wilkinson)
Honda NSX (Long)
Jaguar, The Rise Of (Price)
Jaguar XJ-S (Long)
Jeep CJ (Ackerson)
Jeep Wrangler (Ackerson)
Karmann-Ghia Coupé & Convertible (Bobbitt)
Lamborghini Miura Bible, The (Sackey)
Lambretta Bible, The (Davies)
Lancia 037 (Collins)
Lancia Delta HF Integrale (Blaettel & Wagner)
Land Rover, The Half-ton Military (Cook)
Laverda Twins & Triples Bible 1968-1986 (Falloon)
Lea-Francis Story, The (Price)
Lexus Story, The (Long)
little book of smart, the (Jackson)
Lola – The Illustrated History (1957-1977) (Starkey)
Lola – All The Sports Racing & Single-seater Racing Cars 1978-1997 (Starkey)
Lola T70 – The Racing History & Individual Chassis Record 4th Edition (Starkey)
Lotus 49 (Oliver)
Marketingmobiles, The Wonderful Wacky World Of (Hale)
Mazda MX-5/Miata 1.6 Enthusiast's Workshop Manual (Grainger & Shoemark)
Mazda MX-5/Miata 1.8 Enthusiast's Workshop Manual (Grainger & Shoemark)
Mazda MX-5 Miata: The Book Of The World's Favourite Sportscar (Long)
Mazda MX-5 Miata Roadster (Long)
MGA (Price Williams)
MGB & MGB GT– Expert Guide (Auto-doc Series) (Williams)
MGB Electrical Systems (Astley)
Micro Caravans (Jenkinson)
Micro Trucks (Mort)
Microcars At Large! (Quellin)
Mini Cooper – The Real Thing! (Tipler)
Mitsubishi Lancer Evo, The Road Car & WRC Story (Long)
Montlhéry, The Story Of The Paris Autodrome (Boddy)
Morgan Maverick (Lawrence)
Morris Minor, 60 Years On The Road (Newell)
Moto Guzzi Sport & Le Mans Bible (Falloon)
Motor Movies – The Posters! (Veysey)
Motor Racing – Reflections Of A Lost Era (Carter)
Motorcycle Apprentice (Cakebread)
Motorcycle Road & Racing Chassis Designs (Noakes)
Motorhomes, The Illustrated History (Jenkinson)
Motorsport In colour, 1950s (Wainwright)
Nissan 300ZX & 350Z – The Z-Car Story (Long)
Off-Road Giants! – Heroes of 1960s Motorcycle Sport (Westlake)
Pass The Theory And Practical Driving Tests (Gibson & Hoole)
Peking To Paris 2007 (Young)
Plastic Toy Cars Of The 1950s & 1960s (Ralston)
Pontiac Firebird (Cranswick)
Porsche Boxster (Long)
Porsche 964, 993 & 996 Data Plate Code Breaker (Streather)
Porsche 356 (2nd Edition) (Long)
Porsche 911 Carrera – The Last Of The Evolution (Corlett)
Porsche 911R, RS & RSR, 4th Edition (Starkey)
Porsche 911 – The Definitive History 1963-1971 (Long)
Porsche 911 – The Definitive History 1971-1977 (Long)
Porsche 911 – The Definitive History 1977-1987 (Long)
Porsche 911 – The Definitive History 1987-1997 (Long)
Porsche 911 – The Definitive History 1997-2004 (Long)
Porsche 911SC 'Super Carrera' – The Essential Companion (Streather)
Porsche 914 & 914-6: The Definitive History Of The Road & Competition Cars (Long)
Porsche 924 (Long)
Porsche 944 (Long)
Porsche 993 'King Of Porsche' – The Essential Companion (Streather)
Porsche 996 'Supreme Porsche' – The Essential Companion (Streather)
Porsche Racing Cars – 1953 To 1975 (Long)
Porsche Racing Cars – 1976 On (Long)
Porsche – The Rally Story (Meredith)
Porsche: Three Generations Of Genius (Meredith)
RAC Rally Action! (Gardiner)
Rallye Sport Fords: The Inside Story (Moreton)
Redman, Jim – 6 Times World Motorcycle Champion: The Autobiography (Redman)
Rolls-Royce Silver Shadow/Bentley T Series Corniche & Camargue Revised & Enlarged Edition (Bobbitt)
Rolls-Royce Silver Spirit, Silver Spur & Bentley Mulsanne 2nd Edition (Bobbitt)
RX-7 – Mazda's Rotary Engine Sportscar (Updated & Revised New Edition) (Long)
Scooters & Microcars, The A-Z Of Popular (Dan)
Scooter Lifestyle (Grainger)
Singer Story: Cars, Commercial Vehicles, Bicycles & Motorcycles (Atkinson)
SM – Citroën's Maserati-engined Supercar (Long & Claverol)
Subaru Impreza: The Road Car And WRC Story (Long)
Supercar, How To Build your own (Thompson)
Taxi! The Story Of The 'London' Taxicab (Bobbitt)
Tinplate Toy Cars Of The 1950s & 1960s (Ralston)
Toyota Celica & Supra, The Book Of Toyota's Sports Coupés (Long)
Toyota MR2 Coupés & Spyders (Long)
Triumph Motorcycles & The Meriden Factory (Hancox)
Triumph Speed Twin & Thunderbird Bible (Woolridge)
Triumph Tiger Cub Bible (Estall)
Triumph Trophy Bible (Woolridge)
Triumph TR6 (Kimberley)
Unraced (Collins)
Velocette Motorcycles – MSS To Thruxton Updated & Revised (Burris)
Virgil Exner – Visioneer: The Official Biography Of Virgil M Exner Designer Extraordinaire (Grist)
Volkswagen Bus Book, The (Bobbitt)
Volkswagen Bus Or Van To Camper, How To Convert (Porter)
Volkswagens Of The World (Glen)
VW Beetle Cabriolet (Bobbitt)
VW Beetle – The Car Of The 20th Century (Copping)
VW Bus – 40 Years Of Splitties, Bays & Wedges (Copping)
VW Bus Book, The (Bobbitt)
VW Golf: Five Generations Of Fun (Copping & Cservenka)
VW – The Air-cooled Era (Copping)
VW T5 Camper Conversion Manual (Porter)
VW Campers (Copping)
Works Minis, The Last (Purves & Brenchley)
Works Rally Mechanic (Moylan)

First published in July 2008 by Veloce Publishing Limited, 33 Trinity Street, Dorchester DT1 1TT, England. Fax 01305 268864/e-mail info@veloce.co.uk/web www.veloce.co.uk or www.velocebooks.com.
ISBN: 978-1-84584-162-1/UPC: 6-36847-04162-5

© Graham Templeman and Veloce Publishing 2008. All rights reserved. With the exception of quoting brief passages for the purpose of review, no part of this publication may be recorded, reproduced or transmitted by any means, including photocopying, without the written permission of Veloce Publishing Ltd. Throughout this book logos, model names and designations, etc, have been used for the purposes of identification, illustration and decoration. Such names are the property of the trademark holder as this is not an official publication.

Readers with ideas for automotive books, or books on other transport or related hobby subjects, are invited to write to the editorial director of Veloce Publishing at the above address.
British Library Cataloguing in Publication Data - A catalogue record for this book is available from the British Library. Typesetting, design and page make-up all by Veloce Publishing Ltd on Apple Mac.Printed in India by Replika Press.

SpeedPro Series

THE COMPETITION CAR
DATA
LOGGING
MANUAL

GRAHAM TEMPLEMAN

VELOCE PUBLISHING
THE PUBLISHER OF FINE AUTOMOTIVE BOOKS

Veloce SpeedPro books -

978-1-903706-92-3

978-1-901295-26-9

978-1-845840-23-5

978-1-845840-19-8

978-1-845840-21-1

978-1-903706-59-6

978-1-903706-76-3

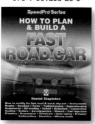

978-1-904788-78-2

978-1-903706-78-7

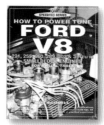

978-1-903706-72-5

978-1-903706-94-7

978-1-845840-06-8

978-1-903706-91-6

978-1-845840-05-1

978-1-903706-77-0

978-1-84584-187-4

978-1-904788-93-5

978-1-84584-142-3

978-1-904788-84-3

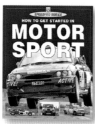

978-1-903706-70-1

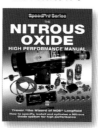

978-1-904788-89-8

978-1-903706-17-6

978-1-903706-73-2

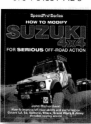

978-1-904788-91-1

978-1-845840-73-0

978-1-904788-22-5

978-1-903706-80-0

978-1-903706-68-8

978-1-845840-45-7

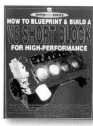

978-1-874105-70-1

978-1-903706-14-5

978-1-903706-99-2

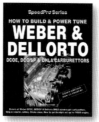

978-1-903706-75-6

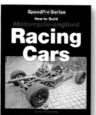

978-1-845841-23-2

978-1-845841-62-1

– more on the way!

Contents

Introduction

If you've got as far as reading the introduction to this book, then you must suspect that logging can help your efforts. You're probably not yet convinced that there's anything in it for you, though, and you might also be worried about whether you'll be able to handle the hardware and software sides successfully. In fact, a common first reaction to data logging is to claim to be baffled by all those impenetrably complex lines on the screen and to be convinced that it's not worth making the effort to understand what they're telling you. On the other hand, however, there's the sneaky suspicion that all those professional teams can't be doing it for nothing, and that there must be something in it. The aim of this book is to de-mystify data logging and get you to a point where you would be very reluctant to run a car without data.

A car always responds to logic, and if it doesn't, it's the logic, not the car that is wrong. Now that data logging is widely accessible, it's harder to blame our lack of success on lack of information; it has to be down to our inability to analyse what's going on, and to come up with suitable strategies. This book is about the analysis of data. Sadly, it cannot be about what you do next. For that you need to look at texts on race car engineering. Here, if I have done the job properly, you'll find how to choose, install and use a data logging system and learn how to interpret what it tells you.

This means that only a very small proportion of the book tells you how to deal with what the data reveals. You'll not find much on how to change or modify the car to overcome the problems. The book will help you identify the symptoms: the cure is up to you. So if, with the help of this book, you're able to identify the symptoms of understeer when the car turns into a corner, the book will not tell you how to deal with it; that's your job. We'll deal with facts and how to put them into theories, but not the next steps.

This book is intended to take over where the manufacturers' guides leave off. It is a small book, not because data logging is a small subject, but because data logging is a means to an end. Data logging is a tool, and its job is to help you go faster. In writing the book I've tried hard to keep this in mind, and to focus on the things that will pay-off, in terms of performance. There's little point in buying and installing the hardware just to create beautiful charts or interesting and elegant maths channels.

There was a difficult decision to be made about how many sensors to deal with. For a professional team, the decision is simple. If a new sensor will improve performance, and there's money in the bank, it will be bought. A professional team would probably log thirty or more separate parameters and use them to create almost as many maths channels. This adds considerably to the time and effort needed to operate the system, and is why many teams find it worthwhile to employ a full time data engineer.

Amateur teams have to make a cut-off somewhere, and here the cut-off chosen was logging suspension movement. This is because the equipment needed starts with displacement sensors, which are fairly expensive, and for the results to be of any use data needs to be logged at at least 200 samples per second. It is financially and technically demanding. Even this wouldn't tell the whole story, and pretty soon you would want to move on to information about actual (rather than calculated) ride height, suspension loads and tyre temperatures as well. Knowing damper movement opens up a whole range of information about wheel frequencies, dynamic ride heights and, by implication, spring and damper rates, so it was a difficult decision to make. Not logging damper movement means that we don't know precisely what the suspension is up to at any moment in time, and this makes our job more difficult, but it does reflect the finance, time and engineering resources available to the majority of amateur teams.

The focus is the principles of data logging in the hope that the book will not go out of date very quickly. This in turn means that there is not much in the way of product comparisons and detailed explanations of software packages. This will probably infuriate the reader when it becomes obvious that some systems lack essential features and the guilty party is not named. The simple fact is that the systems are constantly evolving and improving, and it would be wrong to comment on things that are very likely to change. What you do get with this book is guidance on how to choose a system that will do what you want. I have also tried to provide information that will help to maximise the benefit from the system you have, and to get a real understanding of what's going on with the car and the driver.

The hierarchy in a professional team is well defined. For each car there will be a driver, an engineer (who might or might not also be the team manager), mechanics, and often, nowadays, a data engineer. Amateur teams will not be so well resourced, but somebody still has to cover all these jobs. The reality is that you will need an extra body to look after the logger and extract the information from it.

Although the majority of people reading the book will be part-time competitors with a team consisting of a collection of friends and helpers, I've stuck rigidly to the idea of proper team roles. I talk about the team manager, the race engineer, data engineer, mechanics and driver as if they were separate people, where I know that, in reality, these roles will be shared between you in the way that best suits. Using the idea of roles indicates the nature of the job that needs to be done, not that without these people, you cannot possibly succeed, and it provides me with a useful shorthand.

Data logging is an expensive business but how much you spend depends on your budget and resources. For the well-heeled teams, it's just another operating expense to be accounted for in the preparation of the season's budget. For the less affluent, there are choices to be made. Some will opt for expensive systems complete with looms and sensors provided by the company that supplied the logger, and the justification will be that although the initial outlay is painful, the reward comes in time saved and reliability of operation.

Further down the food chain, money can be saved and sensors can be adapted to provide data that might otherwise be out of reach financially. Cheaper systems are not necessarily inferior systems, but might not be as 'turn-key'. It may well be that the hours spent adapting a sensor would have been better spent working overtime and spending the proceeds on professional kit, but those of us brought up on Meccano and Lego Technic find a certain satisfaction in the do-it-yourself approach. The book deals with both the 'plug-and-play' and home-brewed approaches.

Stylistically, there are a few things that ought to be mentioned. There is always the he/she conundrum in what is undoubtedly a male dominated world. In the case of the book, however, it is not an affectation; much of the data here records the exploits of a lady driver.

For the statisticians who read this book, I know that we are talking about these data, but I shall call it this data. I apologise to engineers for the horrific mixture of units. Like a great many English people, I have a pretty schizophrenic attitude to Imperial and SI units. I measure speed in miles per hour, but do calculations in metres per second. Distances of less than a millimetre are measured in thousandths of an inch, although all other sizes tend to be in millimetres. Most engineering calculations are done using the SI system, but power is bhp (the James Watt, 33,000 pounds/feet/minute variety), and torque for both engines and fasteners is measured in lb/ft rather than Newton/metres.

There are many examples of real data throughout the book. Lots of these do not have the scale shown on the y axis. This is partly to protect confidentiality, but mainly because, for the purpose of this book, the numbers are not important, but the shape and direction of the traces are. Much of the data comes from relatively low-powered sports racing cars, with a very small amount of downforce and racing tyres. They achieve lateral acceleration of about 1.6g. If you race a production car or compete on loose surfaces, reaching these figures is unlikely. If you sprint autocross or hill-climb a purpose-built vehicle, the numbers will seem puny. In all cases the shapes stay roughly the same.

Knowing that the book will be sold on both sides of the Atlantic and in many other places in the English-speaking world, I tried initially to use a mid-Atlantic dialect and failed miserably. It has taken me a number of years of sheer frustration to train my word processor to spell organise with an s and not a z, and there is no way I would let my computer go back to the American spellings for even a short period. The technical terms I use are British English, and wherever possible I have translated into other versions. For example, I use rpm where I would normally talk of engine revs. There are some terms that cause concern. I have watched enough American movies to know that the pedal on the right is the gas pedal, but to me it is the throttle. Over here we talk about ECUs for engine control units, but the language is still evolving and manufacturers use a range of acronyms for the computer that controls the engine and, increasingly, related systems. So forgive me if I've not done a complete job. It's not just that spanners become wrenches when they cross the Atlantic, but that even 150 miles north of where this is being written, older mechanics don't use spanners, they use keys. So, if I can't translate between Leicestershire and Lancashire, be patient with me if what I write doesn't read well at Sears Point, Phillip Island or Taupo.

Let's get started ...

Graham Templeman

Chapter 1
Choosing a system – hardware & software

Modern technology is so complex, and the differences between products so subtle, that it sometimes seems that you have to buy something and use it so that you can learn what you should have bought in the first place. In the case of data logging, success depends on both the hardware and the software, and this chapter looks at the features that are available. It's tempting to be dazzled by the equipment and forget the fact that the real work that you do will be with the software. When this book was written there was one logger that was let down so badly by its software package that it wasn't worth buying. If this chapter works, it will mean that you only have to spend your money once.

HARDWARE

It all depends on your budget, but this is definitely a case where more is better, and you want as big a set of numbers as you can afford. Try to buy as many channels as you can, with the highest ADC bit count and the highest sampling

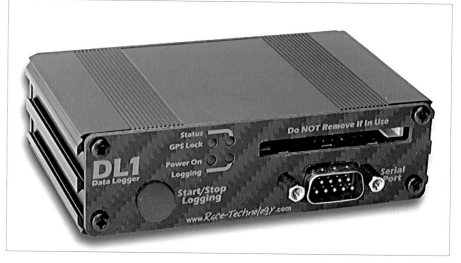

1.01. This is the Race-Technology DL1 which was used to gather much of the data in this book.

frequency that you can afford, and with the most memory.

These terms need explanation. The number of channels is fairly straightforward. The absolute minimum is speed, rpm, lateral g (cornering force), steering wheel and throttle position. These will tell you how the car and

the driver are behaving. Longitudinal g (acceleration and braking) can be calculated from the speed channel, but it's much better measured with a physical sensor, so that's another one.

If you worry about the engine, adding in a range of temperatures and pressures would be good – so that's oil and water temperature, and oil and fuel pressure. There's no doubt that being able to measure wheel speed on all four wheels has its benefits, and measuring suspension movement would add to the understanding of the handling. Throw in another four channels for laser ride height at each corner, perhaps four more for tyre temperatures and a couple for brake pressures, and we're getting towards a channel count much higher than the typical entry level.

In reality, the first five will tell you what you need to know, and beginners should probably stick at that. However, the engine vital signs don't take much effort, and can be interesting, so let's include those as well. So, buy as many channels as you can afford and don't be embarrassed by having some redundant ones at the beginning. Some systems are expandable and, if you buy eight channels to start with, it's often possible to buy extra boxes for extra channels when you need them. So, there's another question for your consumer research.

The ADC bit-count refers to the process of Analogue to Digital Conversion (hence ADC) and most of the things that you log will require conversion from an analogue signal (such as the voltage output from a throttle position sensor) into a digital form that the computer can understand. The number of bits (the bit count) of the ADC is a good measure of the precision with which the logger can operate. Chapter 2 deals with this in more detail but, basically, the bigger the number the better.

The computer's digital logic uses binary arithmetic, and so needs numbers in the form of ones and zeros. The decimal number 9 is expressed as 1001 in binary notation. This is known as a four bit number because it takes four combinations of 1 and 0 to express it. The highest binary number using four bits would be 1111 and would

represent 15 on the conventional decimal basis. So a four bit binary number can represent anything between 0 and 15 and can provide us with 16 individual values. An eight bit number counts up to 255 to give 256 values and a ten bit number would provide 1024 separate values. The bit count gives us an idea of the resolution of the system.

We now have a view of how accurate the measurements can be. So we would be happy to measure throttle position with 8 bit accuracy because it would split the throttle movement into 256 separate parts. If we use a scale of 0 to 100%, an 8-bit ADC would be capable of measuring to 0.39% of the throttle movement (100% ÷ 256) and is more than adequate for understanding what is happening.

For reasons we need not go into here, it's sensible to regard the logger as only half as accurate as its resolution indicates. We would do well to regard the accuracy of the throttle movement not as 0.39% but 0.78% and this is not really significant. In other instances this can matter, and using a 10 bit ADC provides a definite improvement in resolution and accuracy. The enhanced precision can

1.02. Lateral acceleration logged at 100Hz (above) and 10Hz.

be very useful when it comes to installing and calibrating some other types of sensor. The sampling frequency tells us how many times per second each value is read, and you should look for a high number here as well. Most data can be happily collected at ten or twenty times per second (10 or 20 hertz) but some things need greater frequency. If you're trying to measure suspension movement, 200Hz is often regarded as the minimum; so look for the ability to use high sampling rates if you think you might need them eventually. There is a trade-off here. High logging rates give more information when you zoom in on the data, but they use up more memory, and the trace is dirtier to look at. Chart 1.02 shows lateral acceleration logged at 100Hz (samples per second) and 10Hz. The two lateral g traces shown in Chart 1.02 were actually gathered using a Race Technology DL1 at 100Hz and the upper trace is shown at this rate; the lower trace is shown at 10Hz.

The maximum frequency that the logger can gather samples is important, but so is the maximum number of samples per second that can be gathered. 8 channels, logged at 20Hz represents 160 samples per second – well within the range of most systems. But if you decide that you want to expand the system by adding on extra channels and you want to log something at a high frequency (say 4 wheel speeds at 100Hz), then you may be in danger of exceeding the maximum number of samples per second. It's another question to ask and an issue that you might want to future-proof against.

This leads on to memory capacity. Even though you specialise in 10 lap races or short rally stages, when the occasion arises, you want to be able to record data from the longer stints. It's a major pain to have to be careful not to run out of memory. The most popular form is removable memory, such as memory cards, which lets you get the data out of the car and on to the computer quickly and easily and not to get in the mechanic's way. If your system has removable memory cards it's worthwhile having three cards. Whenever a card is taken out of the logger, a replacement is put in. In this way you never lose a session because you forgot to replace the card after loading data on to the laptop. The third is kept in the glove box of the transporter as a backup.

If the logger on your shortlist has internal memory, the question to ask the vendor is not how big the memory is in terms of megabytes, but in terms of time available for logging. Some loggers are much more memory efficient than others. You won't get a straight answer because it will depend on how many channels you're logging and what sampling rate you're using, but look for a time that exceeds the longest race that you expect to do, that covers the maximum number of channels you expect to log, and at the highest rate.

These devices will have a hard life in a hostile environment. Weather-proofing is important so ask questions about the level of sealing. It's not just the open cockpit cars that are open to the weather, closed competition cars are seldom as well sealed as road cars. Check whether the logger of your choice meets the IP ratings (Appendix 1 lays out what the standards are so that you can see what the specs mean). The electrical connections and quality of the wiring is also important. Moisture can be a problem, but so can abrasion and ham-fisted mechanics.

Some loggers use formal plug and socket systems and others provide terminal strips into which the sensor wires can be plugged. Proper connectors are robust but can suffer from clumsy handling, pins can be pushed back into the body of the connector, and water can still be a problem. Screw- or snap-in connectors are much more adaptable if you're of an experimental frame of mind, but if they come loose, you only have yourself to blame. Increasingly, the marketing strategy is to make the systems 'plug-and-play', with the minimal amount of wiring or calibration, which might not suit the experimenter.

REVIEWING THE MARKET
Now that you have an understanding of the sort of hardware that you're looking for, you'll need to conduct your own review of the available systems. As with any other type of electronic products, prices fall and feature levels increase continuously, so what's true at the time of writing is unlikely to remain so for very long. If you couple this with the increasing sophistication of the vehicle's own computer systems and communication networks, it makes keeping up to date an impossible task.

Happily, the movement in prices and sophistication is all in the right direction (from the competitor's viewpoint) and since the principles remain the same, future competitors can expect to have access to the sorts of facilities and features that were at one time the preserve of the very well financed professional teams. Not only does the hardware get better, so does the software. and the competitive rivalry between manufacturers means that when a new feature is introduced by one company, it's quickly copied by the rivals and the user benefits. What we now think of as a very sophisticated system will be regarded as pretty much entry level in a few years' time.

Loggers and dash loggers
The market is currently split between loggers and dash loggers. Some units stand alone and do nothing more than record data for downloading. Dash loggers, on the other hand, combine recording data with providing a display for the driver. Care is needed here, because there are dash units on the market that provide no logging facilities or, if they do, they're limited to the inputs that you would normally expect on a dash (speed, temperatures and pressures and rpm). The critical requirement here is the ability to measure lateral acceleration, because without that data we are denied insights into how the car handles.

A dash logger tends to be found in the middle and upper price ranges,

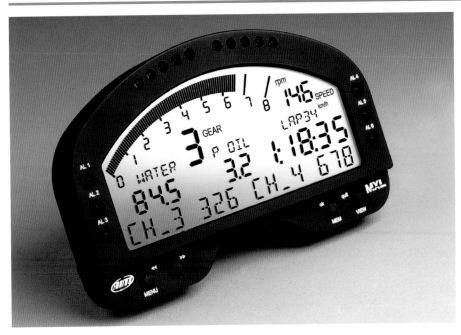

1.03. A typical dash logger – this one logs eleven channels, and has 8Mb of memory. The warning lights and display can be configured to suit the driver.

but can be an economical way forward when the cost of instruments and a lap timing display are taken into account. They should be programmable to display the required rpm range and a choice of information and warnings when parameters stray beyond pre-set limits. Add in the ability to display lap times and provide shift lights and the case for spending more money becomes very strong.

One view is that if data is being logged, the driver actually needs very little information. Warning lights can inform him or her that things are not as they should be, and shift lights will prevent damage to the engine. Add in a lap timer and you could argue there is no need for any sort of dash. Except that some dash loggers can perform one invaluable trick that will probably help the decision. They can predict the lap time of the current lap. They know how far the car had travelled on each second of the fastest lap and can compare that with the distance travelled on the current lap and make a prediction for the

current lap time. Some can even display this as a bar graph of faster or slower. The benefits for the driver are obvious. Not all loggers are equal in this respect, some can only do the calculations for the fastest lap in the current session and others can use a reference lap especially loaded for the purpose. Some can even be programmed to recognise corners and to compare corner times.

At the very top of the range, a dash logger can have onboard maths capabilities so that it can collect data, do some calculations, and present an output. For example, by knowing the amount of fuel that the injectors deliver per second, the actual duty cycle that the engine is operating at, and the rpm, the dash can compute the amount of fuel delivered. It could then take this value away from the starting fuel load and display the amount of fuel or the number of laps remaining.

GPS-based systems

Many loggers use the Global Position System (GPS) satellites to provide data

and draw maps. With some it is an add-on, with others it is the mainstay of the operation, but in either case there are many benefits. A question worth asking is the frequency with which the system updates its fix on the satellites. Quicker is better from the accuracy viewpoint and you should be looking for at least 10Hz.

Not only does GPS reduce the initial purchase and installation costs, but it can also produce better data due to the better method for measuring speed. If the system uses the GPS to update its position, ten times per second, it can calculate the speed at which it is covering the ground. This is in some ways an improvement on the conventional approach which is to measure speed by measuring the time gap between the wheel fixing bolts passing a sensor and doing some calculations based on the rolling radius of the wheel. This 'count the wheel revolutions' method presents us with a few problems. We don't know the rate of tyre growth with speed, so we really cannot be sure that we're measuring correctly, and if the driver locks up the wheel completely it kills the speed signal and distorts comparisons with other laps. Other problems include not being able to log the first second or so of a racing start, and the need to recalibrate the software when changing from one type of tyre to another. But in reality, although all these problems do exist, generally speaking they're more of an inconvenience than a series of major issues.

The only reservation about GPS systems is that people tend to credit them with more accuracy than they sometimes exhibit. GPS systems draw immaculate track maps, and it's fascinating to see the tyre-warming weaving on the green flag lap and the driver taking her place on the grid, but the systems are currently only accurate

1.04. Without GPS, the car will need a beacon receiver.

have GPS or a Pi system, they are an essential part of lap timing and track mapping so you need one. The GPS-based system simply needs you to draw a line on to the map and nominate it as the lap marker and can forget all about beacons and receivers on the car. If you use a Pi system, that company has arranged a 'monster' beacon to be fitted to the start line of most European race circuits. This triggers all Pi systems, and some others as well, but it's as well not to rely on it in case the organisers forget to switch it on. This is unusual except on some test days. That leaves everyone else with the need for a beacon and the chores of putting it out at the start of the day and remembering to collect it at the end of the day. All cars will need a receiver mounted on the car.

to within a few metres, so reviewing racing lines via track maps is a bit speculative. There are ways round this, involving survey-quality systems and triangulation with a separate base station, but by now we're getting into real money again. If the correct line is only a foot wide, a GPS system is not going to show you how to find it. We are promised that the accuracy of the system will improve over time.

A current development is to compare the GPS-derived lateral acceleration data with the onboard data and use this to assess understeer or oversteer. The early signs are promising, and it's certainly easy enough to carry out, so if (when) this turns out to be a robust and reliable method to evaluate the handling, then the case for GPS will be complete. In the meantime, the excellence of the track maps and the convenience of not having to place the timing beacon are pretty persuasive reasons for using a GPS system.

Timing beacons and receivers

Timing beacons are a pain. Unless you

1.05. Putting the beacon on a tripod makes it much easier to aim at the right part of the track.

Relationships with the car's ECU

Another important fact to consider is whether your car is fitted with any sort of management system. Increasingly, engine control units are monitoring many factors around the car and it's likely that the data used by the control unit can also be used by the logger. The current state of the art is the system known as a CAN (Controller Area Network) Bus. This is a set of protocols that provide for sharing data and some loggers can take data directly from the CAN Bus and log whatever is required. This is a painless way collecting signals for wheel speed, rpm and engine vital signs.

The normal physical form of a CAN Bus is a twisted pair of thin wires that carry a 'high' and a 'low' signal, and if your logger is compatible with the CAN Bus, it will be a matter of finding a convenient node and plugging the logger into the bus. The logger then needs to be configured to read the coded messages and use the ones that are of interest. With this system there is instant access to the whole range of parameters within

Data typically available on a CAN Bus		
Engine rpm	Engine temperature	Fuel pressure
Fuel temperature	Barometric pressure	Battery voltage
Inlet air temperature	All 4 wheels speeds	Throttle position
Manifold pressure	Fuel injection duty	Air fuel ratio
rpm limit active	Error states for lambda and injectors	Yaw rate

the system. This is mainly a blessing – you can read rpm, throttle position, temperatures, air fuel ratios, and possibly even wheel speeds and a whole range of other things that are available. An acceptable downside is that you are in danger of information overload especially if you're using whole vehicle systems rather than simply the engine ECU.

Even if the ECU does not use a CAN bus, it's still likely to be monitoring a wide range of parameters, and so it will provide access to various inputs for use in the logger. After-market ECUs that are built by for motorsport are often capable of interfacing directly with logging systems, and from the other perspective, logger manufacturers are aware of the various alternative ECUs on the market and configure their logger to use the inputs to the various models. A key feature of the pre-purchase research is to explore the compatibility of the systems.

If the ECU uses an onboard diagnostic system, such OBD2 or EOBD, this is not as useful as you might hope. Here the data stream is intended for fault finding, each manufacturer speaks its own version of the standard dialect and decoding is difficult. The final possibility is that it might be possible to 'borrow' signals from the ECU by tapping in to the relevant pin on the ECU multiplug. There are certain precautions to be taken with this approach, and these are covered in Chapter 2.

At the expensive end of the market, and with proper interfacing, the logger can even take control of some of the functions of the car's system. This happens when the logger has

onboard maths capabilities, is capable of generating output signals, and is connected in the correct manner to the car's ECU. For example, if wheel speeds are logged, a large variation in the speeds of the driven wheels indicates a traction problem, and this could result in the dash sending a message to the ECU to retard the ignition and provide some form of traction control. This is expensive equipment, and you'll be spending so much money with the manufacturer that they'll be pleased to sit down with you and your credit card to help you decide precisely what you need.

Sensors

If sharing ECU sensors is not a possibility, the manufacturer will be pleased to supply whatever the customer needs. When you buy, it's always worth making sure exactly what you'll get in the package, and what comes as an extra,

because the bottom of any particular range is often not such good value by the time the necessary extra sensors have been added.

You should expect:
• timing devices (infra-red transmitter and receiver)
• lateral accelerometer (to measure cornering force or lateral g)
• speed sensor
• some way of measuring rpm and keeping the signal 'clean'

GPS systems do away with the need for a timing beacon for the pit wall, a receiver for the car, and with speed sensors, although you might still want to buy one. You should still expect to find an accelerometer for measuring lateral g, and a good system will use that and a long g sensor to keep the GPS data honest.

Adding a throttle position sensor (TPS) and a steering sensor will get you really useful information about the handling of the car and the skill of the driver. Forward acceleration and braking (long g) can be calculated using a maths channel so long as the software

Hardware checklist
How many channels?
Does that include any standard ones like rpm and Lateral g?
What sensors are included as standard?
What is the ADC bit rate?
What is the maximum frequency at which data can be logged?
What is the maximum number of samples per second?
How many minutes of logging, and at what number of samples per second?
Is the data lost if the system is switched off before downloading?
Is direct interfacing with the ECU possible, and with which makes?
Is it possible to add extra channels?
Does the kit come with a beacon and a receiver?
If it is GPS based, how quickly does it update?
If it's a dash logger, to what extent are the screens configurable?
Can it predict lap times?
What happens about wiring looms (who supplies them)?
To what extent is it plug-and-play (e.g. is there a list of standard sensors)?

has the derivative function to calculate the change in speed per unit of time. There is a strong case for the physical measurement of acceleration using a second accelerometer. Most setups that only have one accelerometer include long g calculation in the pre-configured maths channels, but direct measurement is better.

If the logger is integrated with a dash display we'll need temperature and pressure sensors unless it's possible to interface with the ECU.

If we're talking about a modern car with anti-lock braking, there are signals available for individual wheel speeds, although most logger manufacturers are anxious about amateurs borrowing signals from safety critical systems.

This has been a pretty long list of things to look out for when buying a logger. The table below summarises the important points.

WHAT TO RECORD?

The range of possibilities is almost infinite and almost anything that the engineers or driver might conceivably be interested in can be measured. The limit is set only by the costs, in terms of money for the system and in time devoted to analysing the information provided. Professional teams tend to log all of the things shown in the accompanying table as a minimum. Amateurs have to live in a world with fewer resources and have to make choices. The more you log, the more it costs, the more there is to go wrong, and the more time is needed analysing the data.

SOFTWARE

In a way, the hardware hardly matters if you hate the software. Luckily, the manufacturers tend to make working copies of their software available for download. If they don't, it should at least raise a doubt in your mind. Do they have something to hide? Are they profit maximisers who are always looking for ways of extracting more money from you through updates or licences?

Playing with a free copy of the software is informative and frustrating. It's good because you can see what you get, and whether you like the look and feel of it. It's bad because it really takes a couple of months of usage to really get into a package and its ways of working, and an evening or two as part of the shopping process doesn't always do the package justice.

'Must-have' features for logging software
Channel reports to show min, max, and average values
Choice of number of variables that can be shown in charts
Control over scaling of variables in charts
Lap and segment timing
Lap times and sector timing that doesn't need extra beacons around the course
Maths channels with logic and derivative functions
Overlays of several laps on top of each other
Strip charts of variables (rpm, speed, etc.) against distance or time
Time slip function to show losses and gains at different points
XY charts to plot one variable against another

Useful software features which you can live without
Choice of whether to show data in 1 or more strips
Control over the height of the strips in strip charts
Creates standard layouts, maths channels, etc., that can be re-used
Deal with sensor offsets by recalibrating graphs (e.g. zeroing the throttle)
Divide laps into sectors automatically
Good range of mathematical and logical operators in the maths package
Histograms to show frequency of channel values
Mixing different types of chart and data (tables, strip charts, XY charts, etc.) on the same screen
No limit to the number of maths channels that can be used
Show corners as shaded areas
Showing data on two axes, but with colours showing a third aspect
Software that translates direct measurement of output into values suitable for display (e.g. volts to degrees of steering)
The ability to show different data in different ways at the same time (e.g. strip chart with speed and lateral g, XY with speed and rpm).

Entry level – essentials	Speed, rpm, lateral g (cornering forces), throttle position, steering input
Entry level – extras	Long g (acceleration and braking), oil and water temperature, oil and fuel pressure, lambda or exhaust gas temperatures, brake pressure (in one or both circuits)
Mid range systems (Entry level plus)	Damper position
Pro-level systems (Mid range plus)	Ride height, air speed, tyre temperatures, force measurement on pushrods or dampers, driveshaft torque, yaw (rate of rotation)

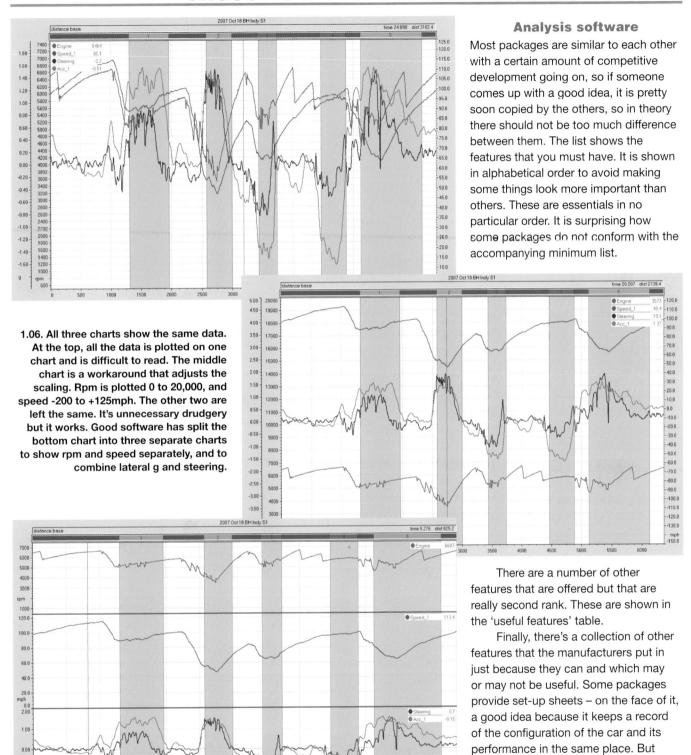

1.06. All three charts show the same data. At the top, all the data is plotted on one chart and is difficult to read. The middle chart is a workaround that adjusts the scaling. Rpm is plotted 0 to 20,000, and speed -200 to +125mph. The other two are left the same. It's unnecessary drudgery but it works. Good software has split the bottom chart into three separate charts to show rpm and speed separately, and to combine lateral g and steering.

Analysis software

Most packages are similar to each other with a certain amount of competitive development going on, so if someone comes up with a good idea, it is pretty soon copied by the others, so in theory there should not be too much difference between them. The list shows the features that you must have. It is shown in alphabetical order to avoid making some things look more important than others. These are essentials in no particular order. It is surprising how some packages do not conform with the accompanying minimum list.

There are a number of other features that are offered but that are really second rank. These are shown in the 'useful features' table.

Finally, there's a collection of other features that the manufacturers put in just because they can and which may or may not be useful. Some packages provide set-up sheets – on the face of it, a good idea because it keeps a record of the configuration of the car and its performance in the same place. But no-one likes using other people's set-up sheets; they never quite hit the spot as regards what you do and how you set

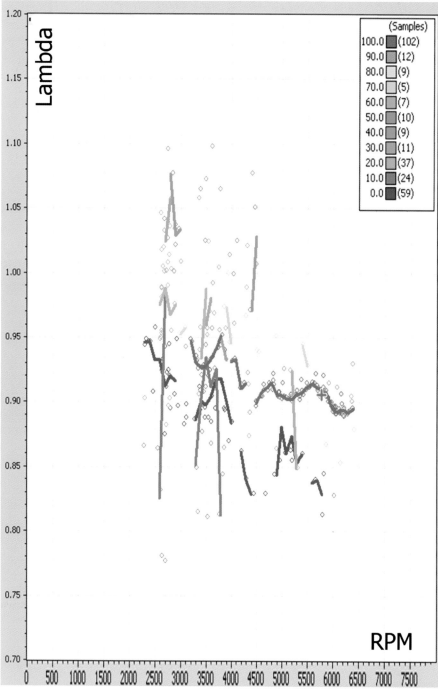

1.07. This chart shows three dimensions; rpm and lambda on the X and Y axis and, by using colour, throttle position.

another showing where time is gained and lost. These animations are useful for entertaining the sponsors, and attracting customers on exhibition stands, but once you begin to understand the software, you're likely to find that you use them less and less.

Software for calibration and monitoring

The logger will come with some sort of setting up software that helps the user to do the basics, like naming the channels and selecting units of measurement that are going to be used. If you're doing a pre-purchase review of software, this is likely to be pretty unsatisfactory because most of these programs need the logger to be connected up before they will do anything interesting. Check out the software anyway, though, just to see how comprehensive it is and what it tells you about how much help that you're likely to get from the manufacturer.

It's another area of choice. At one end of the scale are systems that measure voltages and frequencies but leave you pretty much on your own as to how you go about it. Chapter 2 deals with how to create sensors, and Appendix 2 shows how you might calibrate them, so it's a far from impossible task. The payoff comes in the form of lower cost and greater flexibility. At the other end of the scale, some systems come pretty much completely preconfigured, and the manufacturers actually make a selling point of this. It means that you can use only their sensors and looms and a natural downside here is higher cost and lack of flexibility in choice of sensors. Using preconfigured looms can also be problematical because they're invariably the wrong size to fit your car.

A middle ground is where there is a good range of sensors available from the manufacturer and plenty of support for calibrating them in the system. This

up your car. A notepad facility would be useful, but not many systems provide it.

Most packages will do some form of animation – dashboards which show readings as a dot races round the track map, or how one lap competes against

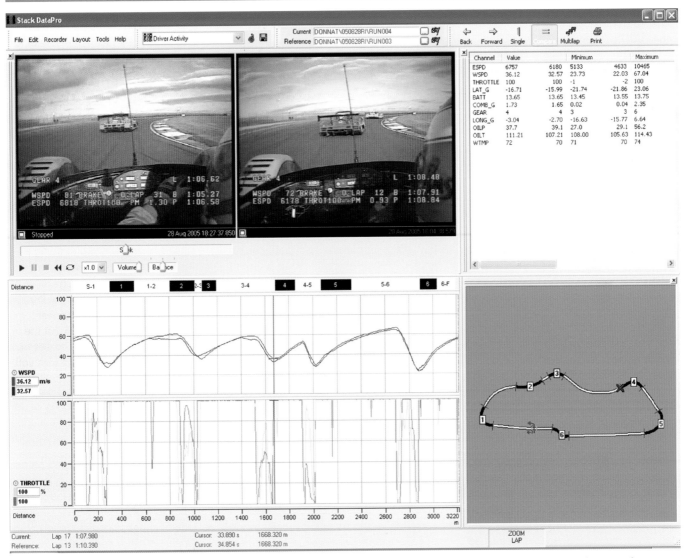

1.08. The two screens allow you to see the effects of traffic, and the impact that the line taken has on cornering speed.

can be in the form of systems that come pre-configured, so that all you need to do is to select your sensors from a drop down list, and the software takes care of the calibration and scaling, and allows you to choose what units you want to work in. At the pre-purchase stage you might want to see that the manufacturer provides calibration formulae that match the sensors that you envisage using with the logger.

Some sensors need extra interfacing, so it's worth asking whether

interfaces are available and at what cost. If you want to use a thermocouple (as opposed to the more usual thermistor-based temperature sensor), it will output a very low voltage which needs to be amplified into the range that the logger can deal with. Similarly, load cells and lambda sensors require special treatment.

VIDEO

At the time of writing, video for logging is a technology that is still maturing. The

computing power needed to link video frames to data points is pretty impressive and, until the way in which computers handle video improves, there is room for scepticism. The problem arises because the generic utility programmes available within the computer industry for handling video clips are aimed at a different market. Their purpose in life is to allow consumers to watch a clip from beginning to end, and possibly do a bit of re-winding and fast forwarding. The data logging demands are different.

Here, the need is to zoom in on a section of data and link it to the video and the software just isn't designed for it. Systems can cope with a bit of data alignment, but need a powerful computer to cope with real time running together of the data and video.

Using video with the data adds another level of complication to the business of running the logger, but it has a lot to offer and is definitely something to look forward to. One major benefit of using video is that you can see what external events affected the data. If the driver takes plenty of kerb, or has to take an unusual line because of another car, the video shows it immediately without having to dredge through the memory banks of what happened when. A video replay is interesting and useful, but far from essential. A team should soon develop the necessary skills to be able to focus in on relevant events and be able to see where the unusual events have occurred and have a good idea what caused them.

The other major benefit is (or will be) the ability to run two videos simultaneously and link it to the data. A small difference in the racing line is difficult to detect using just the data on its own. Show the data together with the video of both laps can show up even tiny differences in the racing line and show the impact on lap times. It will become mainstream and it will be invaluable!

Video is usually an extra cost option involving extra hardware, although you may be able to plug in your existing race cam. If you want to go down this route, you should now have a feel for the sort of questions you need to ask. One cheap option is to use a program that will take independent video clips from a race cam and link it to data from the logger and to display the data at the bottom of the screen. This gives a fascinating insight into the working world of the racing driver, but is time consuming and, if the sole objective is to go faster, not an efficient use of resources.

Finally, a laptop computer provides a suitable means of downloading and storing the data. Any racing laptop will have a hard life. It is at risk from rough handling and the possibility of loss through carelessness or dishonesty, so the temptation is to use a cheap machine. The counter-argument to this is that you will inevitably want to use it to read data and there is nothing more frustrating than a slow machine or a poor screen. A team that has a big enough transporter and a memory-card based logging system might want to consider a desktop machine and a decent screen. Your choice!

Chapter 2
What data loggers do

This chapter deals with how loggers and sensors work. It's not essential reading, but it will help you understand what's going on with the system and why. The chapter will be very useful if you're experimentally-minded, because it will explain how to use cheap and widely available components to measure movement, speed, temperature, and pressure. If your approach is more conservative, and you intend to use the manufacturer-supplied sensors, you'll learn what the sensors are doing, how they're doing it, and you should be able to do the troubleshooting more effectively. It will certainly help you talk to support engineers more confidently when you need to.

HOW DATA ACQUISITION WORKS

Data loggers work by capturing, storing and displaying interesting bits of data about what's happening to the car. A typical system consists of a logging unit which, once switched on, uses its internal clock to tell an onboard microprocessor to read a value from each sensor in turn, and to write the value into memory. This microprocessor has the ability to read a wide range of input information, works at high speeds, and with a high degree of precision.

As an example, the entry-level Race-Technology DL1 is capable of reading fifteen pieces of data one hundred times a second, as well as processing a concurrent stream of GPS data. It converts this data with a resolution of approximately one part in a thousand over the full range. There are, of course, all sorts of ifs and buts in that last statement, which means that a large part of this book is about understanding how much reliance we can place on the data we collect.

The choice of sensors is enormous, but the majority of these will be devices that output a voltage in proportion to the variable that we're measuring. So we can use sensors to measure movement (steering or throttle position, for example)

or temperature or pressure. In all cases, the output is a voltage that the logger reads and presents as a number that the computer can make sense of. 0 to 5V on one channel can mean 0 to 100 per cent throttle opening on one channel and, with a different sensor, the same 0 to 5V can mean 0° to 100° Celsius or 32° to 212° Fahrenheit.

The other important type of signal is a frequency signal from which the logger can measure wheel or engine speed. The sensor generates a pulse every time an event occurs, such as a bolt head passing within range. The logger counts the number of pulses in a set time period and from this it can work out the frequency in terms of pulses per second. What happens is that something like a wheel speed sensor which is pointed at the back of the wheel studs, generates a stream of pulses that the logger evaluates as being, say, 100 pulses per second. The more usual way to express this is as a frequency of 100 hertz (Hz). Since there are four wheel studs each triggering a pulse, the wheel must be turning at 25 revolutions per second. This becomes more interesting if we relate it to the fact that the rolling circumference of the tyre is 1.6 metres (5.25ft). We then know that the speed is 1.6 x 25 = 40 metres per second or 131 feet per second which, in turn, translates into 144km/h or 89mph. How signals are created and how we calibrate them is dealt with later in this chapter.

GPS SYSTEMS

At the time of writing the market share of the GPS-based systems was expanding rapidly, and it looks like they will become a mainstream technology very quickly. The main attribute is the mapping ability (no surprise there) with spin-offs of not needing to use a beacon for track marking, and the ability to measure speed with accuracy.

We do need to be careful (not mistrustful) of the accuracy of the satellite data, because of the rate at which it is collected and the amount of smoothing and processing to which it is subject before we ever see the figures. The accuracy of the systems depends on a number of factors, including the way that the software processes the data, the quality of the receiving equipment, and how well the antenna can see the satellites.

In absolute terms the system should locate our position on earth to within a couple of metres (6ft), and for the bulk of its readings, much more closely than that. But there is a tendency for the readings to drift over time. Over the course of a day, this can show up as maps that don't directly superimpose one upon the other. At the current state of the art, the maps are fine for navigation purposes, but you should be reluctant to trust what they tell you about the precise racing line that you're following. For this you need to be certain to within a few inches. The authorities

promise us that the accuracy will improve.

GPS speed is calculated by measuring the change in position between readings and this seems to be fairly successful but it can be at odds with what the wheel speed sensors tell us. In Chart 2.01 there's evidence of a time lag with the GPS-based system when compared with the traditional system. This shows up most significantly under braking, where the actual error is apparently up to 8mph. On the other hand, the chart shows that there are also calibration problems with the traditional system that are probably caused by poor measurement of wheel circumference. This is obviously wrong by a small percentage because the error gets bigger as the speeds increase. Having both measures gives us a better idea of what's going on.

HOW SIGNALS ARE CREATED

Since the logger will read either voltages or frequencies, we need to look at how the voltages or frequencies can be generated. A variable voltage can be created by using a simple variable resistor configured as a potential divider. This brings some new terms into our language. A variable resistor is also called a potentiometer, often shortened to 'pot'. People also talk about rotary pots, linear pots (where the wiper slides up and down the resistive element

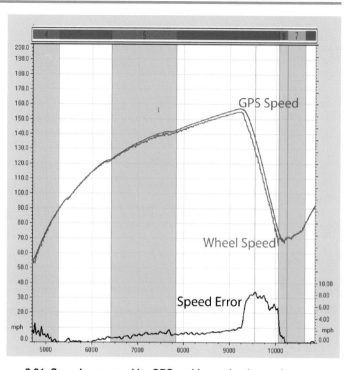

2.01. Speed measured by GPS and by a wheel speed sensor.

instead of rotating round it). They are all variable resistors. Another term for the potential divider is a voltage divider.

Diagram 2.02 shows roughly what you would find inside a rotary variable resistor if you dismantled it. The red and black terminals are at each end of a resistive element which is shown in the diagram in blue. They carry the supply voltage – in the diagram it is 5 volts. The value of the resistance is not important, but is usually 5000 or 10,000 ohms. Running along the resistive element and turned by a central shaft is a wiper. This is shown in yellow on the diagram and it connects the third terminal to the resistive element. If you place the wiper exactly at the halfway position and connect a voltmeter to the zero volts terminal and the wiper terminal, you will see a reading of half of the supply voltage (2.5V in this case). As the wiper is moved along the element, the voltage between the two terminals will vary depending on how far round the wiper is.

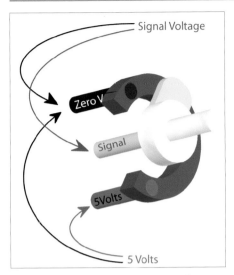

2.02. Schematic of a variable resistor configured as a potential (or voltage) divider. This is the main way in which data loggers measure movement.

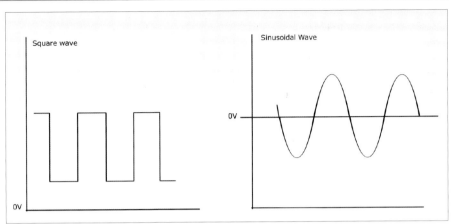

2.03. Loggers generally require a square waveform to trigger frequency measurement.

If we now attach this device to some interesting part of the car (steering or throttle, for example) we can measure the movement and we have created an analogue signal that the data logger can read and record. We would still need to instruct the software precisely what each change in voltage means, and this is dealt with in Chapter 3.

Providing homemade sensors for frequency sensing is not so straightforward. Instead of expecting a steady voltage, the logger needs an output that oscillates between high and low voltage. Typically, the low voltage threshold might be 0.5V and the high voltage threshold might be 4.5V. The voltage would change every time a stud passed the sensor. The change in voltage would need to have a square waveform with a specified peak to peak value for the counting to take place. This is shown on the left-hand graph of 2.03. The standard automotive variable reluctance sensor does not do this, but instead provides a sinusoidal waveform that is shown on the right-hand side of 2.03. The sinusoidal form can be used to trigger a small electrical circuit that

changes it into the required square waveform, but this is a complication that most people will want to avoid. This is a process known as pulse-shaping, and the manufacturer will expect you to avoid these problems with waveforms by using its matched sensors in the first place.

If you feel comfortable using Hall-effect integrated circuits, and happy that you can adapt them to provide a voltage that is within the specification, then there's no problem.

ANALOGUE TO DIGITAL CONVERSION

Once we've provided a voltage or frequency for the logger to measure, it has to undergo a process called analogue to digital conversion (ADC). The signals are analogues and need to be converted into digital values to enable them to be handled by the computer, and in this section we'll look at the ADC process, and how to create reliable analogue measurements.

If we configured a variable resistor as a potential divider by connecting +12V to one end of the resistive element and 0V to the other, we could read a variable voltage between the 0V terminal and the wiper terminal (figure 2.02). If we then fixed it to the steering we might end up with a voltage output of 0.1V when the steering was full left, 5.4V in

the straight ahead position and 10.7V at full right steering. This might measure 25° left, straight ahead, and 25° right at the front wheels. The actual voltage level varies from system to system, with most using 5V, some 12V, and some giving you the choice.

The logger then reads these voltages and translates them into numbers. This is why the advice of Chapter 1 was to buy a logger with a high ADC bit rate. With a ten-bit converter our 0V to 12V can be converted into 2^{10} individual data points, which works out as 1024 separate values. So, our steering output of 0.1 to 10.7 volts will have a range of 904 values from 9 to 913. These values are arrived at in the following way

$$0.1 \div 12 \times 1024 = 9$$
$$10.7 \div 12 \times 1024 = 913$$

This set up can resolve the movements of the steering wheel into 894 (913 – 9) parts, which is more than adequate to tell us what the steering is doing. The software will then convert the number into something meaningful. We could display the steering as simply 0 to 12V, and any values less than 5.4V would mean that the driver was steering left, 5.4 volts would be the straight ahead position, and more than 5.4V would be steering to the right. Most software would give us the ability to restate this as 180° left, dead ahead and 180° right to indicate how far the steering wheel had been turned. We might even prefer to measure the movement of the road wheels, and use 20° left, straight ahead, and 20° right, which could form the basis of a handling channel.

This process of understandable values to the voltage and frequency outputs is what calibration is about. Different systems do it in different ways. Some merely need the figures to be measured and the results entered into a lookup table, and others involve a rather more mathematical approach. You might even find yourself using a spreadsheet to find an equation that approximates the actual behaviour. Appendix 2 shows how to do this.

SIGNAL CONDITIONING
The term signal conditioning refers to modifying a signal to get it into the proper form for the logger to read. We've already looked at pulse shaping to modify a waveform, but there are other instances. For example, strain gauges to measure loads and thermocouples to measure high temperatures have very low voltage output. The type K thermocouple (an industry standard high temperature measuring device) gives 0.8mV (that's millivolts or 0.008V) at 20°C and 20.6mV at 500°C, and this would just be too low for the logger to resolve adequately. Since a 10 bit logger using 5V would only be able to measure down to 4.9mV, it would only be capable of splitting the 480°C temperature range into:

$(20.6 - 0.8) \div 4.9 = 4$ steps or about 120°C (220°F)

This is not nearly enough for understanding our brake temperatures. The solution to this problem is to amplify

the small voltage provided by the thermocouple to a much larger value. Here the original range of 0-20mV needs to be amplified or multiplied by a factor of 200 to bring it into the 0 to 4V range. This sort of operational amplifier is pretty standard electronic circuitry so, if you must, you can buy the components and construct your own. Most manufacturers list amplifier modules that will do the job for you.

Another signal conditioning case is wide band lambda sensors. These are the only ones worth using for logging purposes and these have their own requirements. They need a power supply for the heating element, a frequency generator to activate the sampling, and an amplifier to bring the signal voltage up to a sensible level. In this case, it's definitely easier to buy the proper controller matched to your logger.

SENSORS
For people who are going to buy an off-the-shelf system, there's not much more that they need to know. The company that sells the logger will offer a full range of sensors (complete with the necessary plugs) to fit. The beauty of the ready made approach is that the sensors are ready wired, and pickup their supply voltage direct from the logger and feed back the 0V and signal voltage directly to the correct port of the logger. This saves time and brain-power, but the downside is cost and choice. For the do-it-yourself types and the experimenters, there are a few more things to discuss.

The dedicated low budget operator will realise that good quality variable resistors can be bought new for a few pounds, and a wide range of other sensors can be rescued from an automotive scrapyard. When buying variable resistors, the key is to buy good quality plastic film resistors. The cheaper wire-wound variety lack the accuracy and stability required. If you're buying

your own, check the spec to see that they are plastic film, not wire wound, and buy on price. They tend to come in three ranges: very cheap, (ignore these), twice the price (ok for most applications), and expensive (you may want to use these where precision is important, such as on dampers). The expensive looking plugs are usually commercially available.

POSITION SENSORS
Rotary pots are cheap and useful, and can usually be driven through some form of linkage. Linear pots are also great, but there's no sensible, low-budget, do-it-yourself option. If it absolutely must be a linear sensor, the only ones worth using are the purpose-built linear transducers with proper sealing against the weather, and which can be specified in whatever length you require. There don't seem to be any general-purpose linear pots on the market that are accurate enough electronically and robust enough mechanically to withstand life on board a competition car, so it's a case of biting the bullet and buying good quality specialist items.

If you do decide to go down the homemade route with rotary position sensors, there are some things that you need to be aware of. The operating voltage is not material – the logger will provide you with the necessary power feed at its preferred voltage level. If your logger has screw-type terminals, +5V +12V and 0V should be properly marked. If your logger has a wiring harness with sockets dedicated to analogue signals, there are likely to be only three pins. You need to use a meter to identify which terminal carries the positive voltage and which is the ground or 0V terminal. By default, the remaining terminal will be the signal voltage.

This test will also show you the voltage your logger wants to read. Some use 12V, but most stick to the 5V beloved of electronic engineers. Others

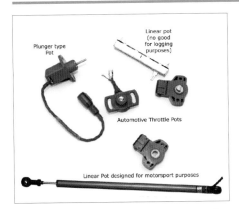

2.04. A collection of potentiometers, only some of which would be robust enough for logging.

give you a choice. Using 5V in a 12V port simply limits the signal you can read (although some software provides for this), but 12V into a 5V port might well damage the unit.

When using rotary pots, care is also needed to make sure that you're using the full scale that they provide. Rotary pots come with varying amounts of travel, with some having a resistive element that is fully covered with a wiper travel of about 110°. These are great for using as throttle position sensors where the throttle only rotates 90°. Others will have nearly a full turn of travel (330°), and they're also available with three turns or even ten turns. So, it's important to match the travel to the application.

You need to be careful about this. On the one hand, you need to deliberately sacrifice some of the available movement to protect the pot. You don't want it to be banging up against the end of its travel, because that's likely to lead to some sort of physical damage. But the margin of safety that you leave eats into the available accuracy

because, instead of measuring using the full 5V scale that the manufacturer envisaged, we might only be using 4V, and the problem is that the inaccuracies can stack up. We lose a bit of accuracy by careless installation, some more by building in the margin of safety, yet more because of the process that converts volts into numbers (ADC) and then a final slice by failing to calibrate the system accurately. We need to be really careful to ensure that we don't create our own version of 'garbage in, garbage out'.

On the other hand, not everything needs to be measured with precision. When you're looking at the throttle trace, all you need is the general shape. Because of the very nature of motorsport, the driver is on the gas or out of it completely. The transition stage is important, but you're more interested in its shape rather than the absolute values.

If you intend to use the logger simply to know what's happening to the car out on track, then not even the steering needs calibrating accurately.

The steering trace for the first turn in Chart 2.05 is indicative of a high speed corner (not much steering lock needed) and the chart shows that the driver was back on the throttle 0.4 seconds after turning the steering. It also shows that she is driving pretty near the limit because, in the second turn, the extra 3mph provoked significant understeer. It doesn't need an analytical genius to see from the steering trace that extra lock had to be applied on the faster run through the corner. Without any formal calibration and measurement we've put together a pretty useful set of insights into the performance of car and driver. On the other hand, if we intend to use the data as the stepping-off point for a deeper engineering analysis of the dynamics of the car, knowing that the actual peak steering angles were 1.7 and 2.5 might be very useful indeed.

TEMPERATURE AND PRESSURE SENSORS

Most automotive temperature and pressure sensors indicate changing

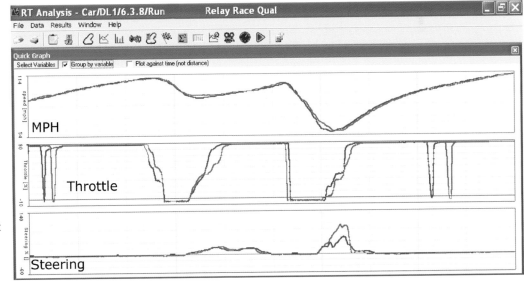

2.05. Careful calibration is not always necessary. Here, the shape of the throttle trace (second from top) shows short sharp spikes downwards, indicating the gearchange, significant hesitancy coming off the gas, and a gentle (but realistic) feeding in of the throttle in the first part of the turn. The extra steering that was needed on the red lap can also be clearly seen.

levels by changing their resistance. The owners of the upmarket systems will simply select the sensor from the list in the software. The experimenters have a bit more work to do.

There are, broadly, two types of resistive sensor. The traditional automotive one has a single terminal tag which earths through its body, and the resistance is between the earth (ground) and the connector. The other type is more modern, and has two pins that typically interface with some sort of control module. In this case, one pin is the supply voltage and the other the signal. If you're using this sort, it will be your job to find which pin is the supply and which is the signal on your car.

When trying to use either type, the first stage is to measure the resistance across the range of conditions. So, if it's temperature you're looking at, start by measuring resistance with the sensor in a jug of ice cubes, then at room temperature, and finally in a jug of hot water. Make a note of these temperatures and the associated resistances. If you're measuring pressure, then you have the slightly more complex job of providing various pressures at which to measure the resistance. Some typical values are shown in the accompanying table.

We can now use the same logic as for the potential divider. If we consider the sensor to be the equivalent of one half of the wiper in diagram 2.02, then all that is needed is another resistor to be wired in as the other half to provide a simple divider circuit that will output a

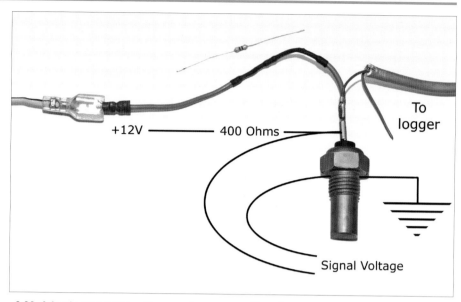

2.06. Adapting a traditional temperature sensor. There's a resistor inside the heat-shrink sleeve, and the blue 0V cable ought to be grounded.

voltage proportional to temperature. 2.06 shows how this can be done.

The value of the resistor can be calculated with reference to the measured values and required range, but a simple rule of thumb is to pick a value that is at about the half-way point in your scale. In the case above, there was a 470 ohm resistor in the toolbox and, when fed with 12V, it gave a voltage range of 1 to 3.5V, adequate for temperature sensing. These voltages then need to be calibrated in the logger set-up software, or using the method shown in Appendix 2. Whatever you do, make sure that the range of voltages that you intend feeding into your logger are within its safety limits. Whether or not you can do this depends on the manufacturer providing a suitable interface.

BORROWED SENSORS

If the car has its own ECU, the easy way to borrow signals is to interface the

logger directly, either through the CAN Bus, or through a serial link of some sort. It's then simply a matter of using the set-up software to pick out the necessary data from the data stream. Even if direct interface to the ECU is not possible, there seems little point in providing a second temperature sensor if there's one there already. We ought to be able to borrow the signal for use by the data logger. The important thing is that the logger must not be robbing the first user of the signal or any part of the signal. The reality is that most loggers can read the voltage from a sensor without having an effect on the host system. The way to check is to measure the voltage seen at the ECU input pin under a certain condition without the logger attached, attach the logger and start logging and then check that the voltage at the ECU has the same reading as before the logger was started. There's more detail about this in Chapter 3.

Temperature	Resistance (ohms)
19°C	1430
20°C	1291
58°C	470
88°C	190
100°C	175

Chapter 3
Installing the system

This chapter looks at how to install the logger and its sensors. When the book was planned, the intention was simply to recommend that you use the manufacturer's manual. However, a sense of insecurity, and enough experience of how things can go wrong, led to the writing of this chapter, which tries to help you avoid some of the pitfalls involved. The best place to start, however, is still the manufacturer's instruction manual, because it's going to provide information that gives the product the best possible chance of being used successfully. Read it and try very hard to do it by the book. You'll also find that this pays dividends later when it comes to setting up the software. Manufacturers might seem a bit fussy, but their instructions are based on long experience of customers' helpline calls.

NOISE, VIBRATION & HEAT

A competition car provides a very harsh operating environment, so all possible steps should be taken to minimise the effects. The enemies are electrical noise (interference), vibration, heat, and the wet. You would expect electrical noise to be a major source of problems, but the manufacturers seem to be pretty successful at dealing with it, so simply avoid siting either the logger or its wiring close to 'noisy' areas, such as the ignition system, generators, and ECUs. It also helps to avoid bundling cables into neat runs because there's the slight danger that signals can interfere with each other.

The heat problem should make you wary of installing in the engine bay, or in any part of the car where there are cooling or exhaust pipes. If you have to do this, take care. Don't forget that integrated circuits cease to work at 50°C (120°F), and it's likely that any manufacturer's guarantee would consider such damage a self-inflicted injury.

Vibration is another difficulty, especially on purpose-built competition cars where the engine is mounted directly to the chassis. What feels like unpleasant harshness to the driver can be a destructive frequency, that just accentuates the stiffness of the rubber mountings and the weight of the unit to shake delicate electronic components to pieces. Anti-vibration mountings have to be chosen very carefully because the mass to be damped is so light that anything too stiff is likely to be of very limited benefit. The chosen damping material should be soft enough for the excited mass of the unit to have some effect on the elastomer. Be careful also that your mounts cannot 'bottom out' by damping in one direction only. The mounting on the left of Figure 3.01 will only damp vibration in one direction. Care must also be taken to ensure that the bolt cannot contact the sides of the mounting holes.

Satisfactory (if somewhat messy) isolators can be made from low modulus silicon sealer sold for use about the house. Nor should you overlook industrial strength Velcro (hook and

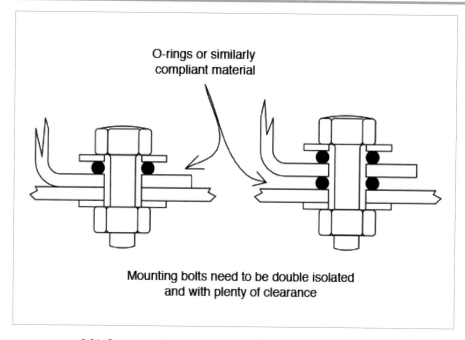

O-rings or similarly compliant material

Mounting bolts need to be double isolated and with plenty of clearance

3.01. Care must be taken when devising anti-vibration mounts.

loop) fastener strip that can provide a convenient way to fix units to the car. It has a useful damping effect that provides a rigid enough mounting without transmitting vibration directly.

The main systems unit should also be kept dry, and this can present difficulties in most forms of competition car. Some loggers are sealed against the ingress of water to quite a high level, and some aren't, but even so, there's no point in taking extra risks. Installing it some distance above the floor will avoid puddles, and there's a lot to be said for installing the unit in a waterproof box. A sandwich container secured somewhere in the cockpit makes a cheap and lightweight waterproof enclosure. It does present problems with routing cables into the box because each will need an entry hole, and this reduces the completeness of the seal.

Mounting the logging unit
Where you mount the logger will depend on several factors. If it's a dash logger, then it's pretty obvious. If there is a

separate unit, then there is a wider choice, and you'll want to balance its physical security (avoiding accidental and environmental damage) with ease of access. The wiring must also be looked after carefully and it must be possible to get at it without problems.

The main unit is likely to contain the accelerometers, and there are some simple rules about where these should be mounted. To measure lateral g, the unit should be mounted as low as possible and near to the longitudinal centre of gravity. To measure longitudinal g the accelerometer should be mounted on the centre line of the car but the position relative to the centre of gravity doesn't matter. These two criteria do not conflict, but the layout of your car might do. They do seem to be counsels of perfection, because some dash loggers have the accelerometer mounted in the dash unit so that you're forced to break the rules. Doing so must have an effect on the accuracy of the logger, but it's an error that is repeated consistently, and only affects comparisons from car-to-car,

and not on one car where all the runs are recorded with the same error.

Any accelerometers should be mounted parallel to the axis that they're going to measure, and horizontal, so a dash containing an accelerometer will need to be at right angles to the centre line of the car, and vertical. One that measures longitudinal acceleration should have its axis in line with the centre line of the car. Units should have their axis indicated by an arrow and, although you can get away with misalignment by adjustments in the software, it makes no sense to build in an inaccuracy. If you can make a mounting platform that has jacking screws on to get things perfectly horizontal, so much the better.

Power supply
The logger needs a power supply and if you want your logger to survive, you must follow the maker's advice about the rating of the fuse to use. You should not need to be reminded to provide for proper terminated connections and to keep the live cable away from damage and the possibility of being severed. If the logger is moved from car-to-car, you need to be sure that any live wire cannot earth against the chassis when the unit is disconnected.

Since electronic noise tends not to be a problem, it makes sense to hook up to the car's power supply first and only go independent if interference proves to be an issue. After all, a small 12V battery (burglar-alarm size) offers a cheap and easy fix. If you do fit a dedicated battery, think about the switching aspect. The master switch will only operate on the car's main circuits, and leaving the logger switched on raises the possibility of a 12V spark igniting fuel vapour, causing a disaster. The circumstances would need to be pretty exceptional, but that's what you usually find in accident investigation reports.

GPS antenna

This needs to be mounted at the top of the car, with a clear view of the sky. Because the standard item is magnetic, the only problem for cars with steel roofs is finding a way to get the cable neatly across to the logger. A composite roof might need a steel panel (inside if the roof is very light, or outside if it has more substance). The steel plate is not just for the benefit of the magnet, but also to provide a ground plane that reflects any unwanted noise from the car's electrical system. It need not be large; you're merely trying to put the antenna into the shadow of the plate. Some people have installed the antenna under the front or rear window, on the assumption that route-finding GPS systems work this way. However, since this inevitably limits the antenna's view of the sky, there seems little point in building in problems and inaccuracies.

With an open car, the roll bar provides the most suitable mounting place. The radius at the top of the bar does not help, so use a small piece of steel sheet as a bracket, and this will also create a ground plane. A few minutes' work with a drill, a hose clip, and a 5mm (or 10-32UNF) button-headed cap screw will make a pretty neat mounting platform. Give the antenna a thin foam rubber cushion if it doesn't already have one.

In all cases, be very careful with the cable from the antenna, because it's fragile and will not stand abuse. It can be fatally crushed, so if you are thinking about routing the cable through a door or a window seal, be sure that it cannot be damaged. It's common sense for a roll-bar mounted unit to use cable ties to keep the cable out of harm's way, but think about the possibility of an accident. The rescue crew will be anxious to get the car out of the way, and in the heat of the moment will probably want to tie a strap around the roll bar. The tightening

3.02. GPS antenna mount. The hole allows screwdriver access, and provides a cable route.

of the strap can sever the antenna cable, so be careful to find a route down the bar that is unlikely to be used as an emergency tow-hook

Using existing sensors

The next logical stage is to go on to look at the installation of sensors. If your logger interfaces with the car's CAN Bus or ECU, then you simply need to pick off the channels in the setting-up software. If you don't have that facility, there might still be the chance to borrow a signal from the car's ECU or dashboard instruments. The first thing to do is to investigate whether borrowing a signal will affect the readings that the ECU or gauge see because there is a very slight possibility that bringing the logger into the circuit might actually do this. If you're worried about the effects on a cockpit gauge, simply add the logger to the circuit and check whether the reading is affected. With an ECU, the thing to do is to measure the voltage normally seen at the appropriate pins of the ECU before connecting the logger. When you know, for example, that a particular temperature reads 3.2V with

the meter across the signal and 0V pins of the ECU, then you can share the signal between the ECU and the logger and make sure that the ECU still sees the 3.2V that it needs to. If the ECU voltage is affected, then you'll need to think about a different signal source for the logger. Generally, loggers have very high impedance and don't interfere with ECUs, but you'll have to judge your own setup.

For things like pressure and temperature sensors, and for throttle position, this is usually not a problem. Getting access to the pins can be difficult, though, so you'll have to break into the wiring at some point. After that it's simply a matter of reassuring yourself that you've not interfered with the value of the signal. If the car started its life as a production vehicle, rpm signals can usually be found as an input pin to the back of the stock tachometer, or as a pin on the ECU output.

Temperature and pressure sensors that are connected to an ECU tend to have a two-pin connection. If this is the case, it's likely that the sensor acts like a potential divider, receives a 5V feed from

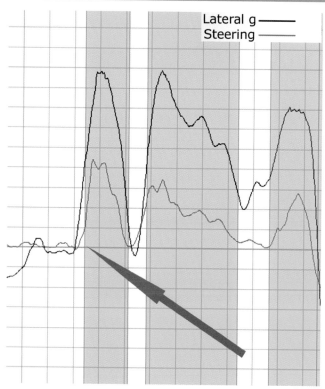

Lateral g ———
Steering ———

3.03. Flimsy brackets can lead to false information. Here, the lateral g builds up before the steering wheel is turned.

3.04. Steering sensor on an adjustable bracket, with the pulley size carefully chosen to give the right range.

choice is extruded aluminium angle or channel with a thickness of 3 to 4mm (⅛-⁵⁄₃₂in). If a bracket breaks it's a pain; if it's flexible it can introduce error into the reading.

The pair of traces shown in figure 3.03 shows an apparently impossible situation. The red line represents the steering, the black one lateral g. The lateral g starts to increase before the steering wheel is moved. This 'miracle' was traced to a flimsy mounting bracket for the linear potentiometer that measured the steering rack movement. The bracket flexed by a few millimetres when the wheel was turned, and once the spring on the bracket had been taken up, the pot started to record steering movement. The data in this case was so obviously wrong that a reason had to be found. In other cases where the mounting of sensors is either flimsy or not properly aligned, it would be possible to overlook the fault and accept incorrect data.

Installing position sensors

The movement that we want to measure will be either rotary or linear, so the choice of which to use seems pretty obvious. Bear in mind, however, that, on a car, most motions get transformed from one to the other at some point. The linear action of the throttle pedal is turned into the rotary action of the butterfly, and the rotation of the steering wheel becomes the linear motion of the

steering rack, so there's often a choice as to what type of sensor to use. A production car with a steering column surrounded by casing and collapsible structures does not lend itself to using the traditional drive belt arrangement and the easier course of action might be a linear sensor of some sort.

In Chapter 2 I pointed out the need to protect the sensor by not crashing into the end stops. The steering pot in Photo 3.06 has a pulley that is carefully chosen to make the best use of the specification of the variable resistor that it's based on. The steering wheel rotates 1.5 turns lock-to-lock. The pot has a full scale of 1.5 turns, so the pulley was made 1.7 times the diameter of the column so that just about all of the available movement is used up when the driver went from full left to full right steering. This means that the full travel of the sensor is not used and a certain amount of accuracy is lost. Instead of reading values from 0 to 12V, the range available becomes 2 to 10V.

Steering

Measuring steering wheel movement always presents challenges, but it's well worth the effort. The linear pot is easy to install and reliable. It sits on the steering rack and is pulled by the track rod. It is also right in harms way. They are expensive and vulnerable – not a good combination. An alternative is to measure steering wheel movement from

the ECU, is grounded through the body of the sensor, and the second pin is the signal for the ECU. The logger should be capable of taking a signal from the signal pin and its own internal ground (0V). You do need to be careful about the physical connections to make sure that they're not going to fail in service.

Sensor mountings

As a general rule, all sensors need to be mounted out of harms way, somewhere where they won't make the mechanic's life harder, and on very rigid brackets. Out of harms way is necessary because Murphy's Law applies. Anything that can be damaged accidentally *will* be damaged.

Folded aluminium just won't do. Any sheet thin enough to bend easily is unlikely to be robust enough to support a sensor. We're looking for rigidity and vibration resistance, so the material of

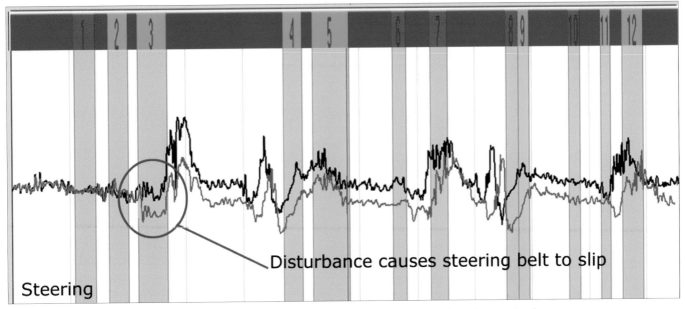

Disturbance causes steering belt to slip

Steering

3.05. If the steering potentiometer drive belt slips, the trace shifts and spoils the data.

the rotation of the steering column. This is easier on purpose-built cars than production cars where the column is surrounded by all manner of stuff. If you can get reasonable access to the column there are several options of which the Hall Effect sensor is top of the range. It simply fits round the column and senses the rotation. There is no contact between the sensor and the column so there are no moving parts to wear out or adjust. It's pretty much a case of fit and forget, until the credit card statement arrives, that is.

Rotary pots can be used. They're mounted on a suitable bracket and driven by a belt of some sort. The simplest form of drive belt is an O-ring, and these can be either bought to size or made in the workshop by cutting and gluing using the correct grade adhesive. A small piece of abrasive paper stuck round the steering column helps the belt grip. This set up is fairly reliable but can suffer from slippage either during operation or in the workshop. If using O-rings, you'll need to make the bracket adjustable so that the O-ring can be tensioned by sliding the pot away from

the column. Be careful not to put the sensor under too much strain because it was not designed to be operated in this way.

Chart 3.05 shows how the data becomes meaningless if the belt does slip. To get reliable and repeatable data time after time, you need to check that the pulleys are aligned properly every time that you run the car. The best plan is to set the steering dead ahead, and mark the two pulleys so that you can see immediately if they have moved relative to each other. Check each time you run the car and, if there's evidence of movement, connect the logger to the laptop and see if it's giving you the required voltage or ADC reading. One more item for the check list.

A better bet is to use very small toothed wheels and belts. This is more expensive and more difficult to set up, but does give consistent results. This can be tensioned in the same way as the O-rings by sliding the sensor away from the column until the belt is under slight tension. Again, heed the warning about straining the bearings of the sensor, or better still, manufacture a bracket that

supports the pulley and leaves the pot as a strain-free passenger. The adjustment is important because done badly the belt can jump the teeth on the wheels. You should also be aware that it is possible for small stones or debris to interfere with the smooth operation.

The best method is probably the one that you will eventually end up with anyway. If at all possible, the pot should be driven directly off the back of the steering pinion. This will need some

3.06. Toothed belts are a good way of driving the steering sensor.

machining on the pinion housing and on the pinion shaft itself, but the end product of all this is a simple, permanent mounting that will not lose its setting.

Throttle position sensor

The throttle position sensor also has a hard life, and some teams reckon on replacing theirs every season. If the car has a tps to drive the fly-by-wire system, or simply to provide information for the fuel injection or ignition, you should try to use it. If you have to provide your own, there are three possible locations: at the pedal, on the cable, or on the throttle spindle.

tps will have to be re-set. The plunger type pot has to be carefully positioned so that its full travel is used up, and this can mean installing the pot a certain distance up the pedal which is not always easy to do. The pedal itself is often not particularly rigid, and when the driver starts jumping up and down on it, it can stray from its original arc of movement. Any slight misalignment can bend the delicate spindle of a plunger type pot and cause it to stick, ruining the data.

At the other end of the system, the pot can be driven directly by the spindle of the carburettor or fuel

it absolutely accurately with the butterfly spindle, it should survive.

Photo 3.07 shows a well mounted, rotary sensor that came originally from a junk yard. The key features are that the bracket is stiff enough to ensure that there is no unwanted movement through flexing, and that there is plenty of provision for adjustment. The slots that allow the bracket to be slid towards the spindle can be seen, the potentiometer itself has a range of adjustment available (the smaller curved slots) and the central register that the sensor fits in is deliberately oversize so that the sensor can be lined up perfectly with the spindle. The thing is a few seasons old and, although the yellow heat shrink casing was a fashion mistake, the epoxy that was used to support the cables as they exit the body of the potentiometer, and the careful attention to alignment, have helped it to survive several seasons. This is proof that taking care can pay dividends

Another possibility is to operate a rotary pot by attaching a lever to the cable so that when the cable is pulled, the pot is moved. If this can be arranged it is simple and reliable. It is better done at the engine end of the system. Photo 3.08 shows about the simplest mount possible. If the bolts were insulated with some soft rubber, it would stand a good chance of surviving.

3.07. Mounting throttle position sensor.

Whatever method is chosen, there must be no chance of causing the throttle to jam open. Bear this in mind when you devise the mounting, and test it thoroughly before you use it. Don't let the car out until you are completely happy with it.

The pedal is not the ideal location because on most cars it's pretty well inaccessible, and because the linkage gets adjusted from time to time so the

injection butterflies. This does not have an easy life either. It lives very close to the engine, so is subject to heat and vibration, but the main thing that seems to kill a tps is misalignment. They often seem to last for no more than a single season on a racing car, and yet the same sensor on a road car would be expected to be reliable for at least 150,000 miles. So, if you use a proper automotive sensor and take care to align

3.08. A simple and adequate mounting for the throttle position pot.

Wheel speed sensors

Unless you use a GPS-based system, you'll need at least one wheel speed sensor to count pulses from the wheel. Sometimes these pulses are from a specially-made 'chopper wheel' that is attached to the back of the hub and in other instances the sensor simply looks at the wheel studs or mounting pegs. In all cases the sensors should be mounted on the non-driven wheels and, in the case of a single sensor system, the outside wheel is the one to choose. So, if your tracks are predominately counter-clockwise, the right-hand wheel is the one to use.

A good tip for accurately aligning the sensor with a wheel stud or locating dowel is to manufacture the bracket and attach it to the upright. Remove one of the wheel studs or locating dowels from the hub and rotate the hub until you can see the middle of the bracket through the hole. Use a drill that is a snug fit in the hole just to mark the bracket and then remove the bracket and finish the job on the drill press.

you must stick to the manufacturer's recommendation absolutely. This is easy enough if the body of the sensor is threaded, but will need care with making the bracket and shimming the sensor to the correct clearance. One sensor will get the job done, two will be better when one wheel locks up (program the logger to display the higher of the two readings).

With only one sensor, a locked wheel can cause a sharp downward spike in the speed trace, and also has knock-on effects. The long g channel, which, in this case,

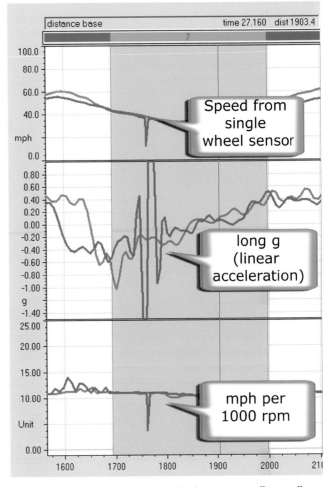

3.10. In this fragment, the wheel locks momentarily, sending the calculated long g trace crazy, and affecting the mph/1000rpm trace as well.

is calculated from speed rather than measured by an accelerometer, goes wild for a short time, and the mph/1000rpm which is also calculated (using speed and rpm) is also affected. These are easily explained and relatively trivial effects, but had the event lasted for more than the 0.2 seconds that it did, it could have affected the accuracy of any map that was prepared from the data, and the ability to overlay one lap's worth of data onto another. If two sensors are fitted, not only can you see which wheel locks up first under braking, you can read the speed from whichever of the two wheels is still revolving.

Measuring temperature and pressure

You should take care to protect engine-mounted temperature and pressure sensors. Many competition engines have either solid or stiffer engine mountings and this can lead to damaging vibrations.

3.09. Some wheel speed sensors are easier to mount than others.

Different types of sensor have different requirements for the clearance between the sensor head and whatever it is that generates the signal, and

To avoid this, connect the pressure sensor to the engine with a short length of braided hose. Then mount it somewhere on the chassis or body with some form of shock isolation. You will probably not do this until one or two have failed in service. Temperature sensors should be mounted in an adaptor that is let into a length of hose, and in this way the natural shock-absorbency of the flexible hose will protect the sensor.

Cables and wiring looms

The 'plug-and-play brigade' will feel pretty smug when it comes to connecting up the various components of the system. Their only problem will be that a 'one size fits all' wiring loom will inevitably have more cable than is needed to connect the various components together, and the surplus will have to be hidden wherever possible.

Providing your own loom can result in a much neater installation. It's not complex, but it is slightly different to the normal wiring tasks. Properly done, signal cables should be the stuff called shielded twisted pair, but in all honesty it is not necessary. The clue is that the sensitive items (like the rpm sensor) usually come with shielded wiring if it is needed. Twisting two normal cables together can reduce the chances of interference, but again it is not really necessary. If you want to do it trap one set of ends of your cable in a vice and use a low-powered cordless drill to do the twisting. More important is to take care to route the cables away from electrically noisy areas of the car, such as ignition systems, alternators, and microprocessors. It's best not to truss things up too neatly, because the chance of interference between the signals increases in large bundles of cable. It's

often better to allow each cable to follow its own route to the logger to reduce this possibility.

The cables also need physical protection because they are vulnerable. Even cables tied neatly out of the way on suspension members can find themselves in the line of fire from flying rocks and debris, and from other suspension pieces that get displaced as a result of the accident. Braided hoses can chafe through cables, especially if the braid is beginning to wear. Use heat-insulating sleeving where cables need to be protected from excessive heat.

If this whole chapter seems a bit over the top, just imagine what it feels like to lose some or all of the data because of careless installation. This is definitely a case of do the job properly so that you only have to do it once.

Chapter 4
The standard data channels

Now that the logger is installed and running, it's time to look at what we are measuring. This chapter deals with the standard data channels and how to extend the channels using mathematical and logical functions. The data in the chapter is all presented in the form of strip charts. Chapter 5 looks at helpful ways to present the data to help us get deeper meaning from it, and Chapter 6 deals with ways of making sense of the vast amount of information that you have collected.

THE STANDARD CHANNELS

The amount of information that the logger provides us with will be pretty much dictated by how much we spend (to buy lots of channels) and how much effort we make (to use alternative approaches to acquire data). But whatever the budget, and however much extra effort you devote to logging, there are a number of key channels that are absolutely essential to understanding the behaviour of the car. They are:

- Speed
- rpm
- Lateral g
- Steering
- Throttle

Each has its own distinctive signature, and it won't be long before you can recognize the main data traces simply from their shape. You'll rarely need to refer to the label to be able to identify what it is you're looking at because of this characteristic form. The speed trace climbs disappointingly slowly but falls away very quickly. Rpm is much the same but has a saw-tooth pattern that indicates the gear changes. Lateral g oscillates around zero along the straights and climbs and falls through the corners. Steering follows a similar pattern, but the movement is much less pronounced in slower corners. The throttle spends long periods at its maximum reading and spikes down for gearshifts and braking areas.

The strip charts here are all measured against distance, and the software is used to identify corners and colour them a tasteful shade of cream. The coloured banner across the top indicates straights as blue, right turns as green and left turns as red. This eases the process of navigating along the trace, since it simply becomes a matter of counting the corners to know where you are in that manner. Sadly, it destroys nearly a century of motor racing tradition when corners like Paddock Hill at Brands Hatch lose their individuality, and become simply the Turn 1 of the strip chart shown here. Somehow, Turns 2 and 3 at Spa don't sound half as challenging as Eau Rouge. But is also avoids miscommunication, so if people in the team don't know their Esses from their Elbows it has no impact on the data analysis.

SPEED

The speed channel is ultimately the only one that we're interested in. It is the objective of our efforts, and all the others

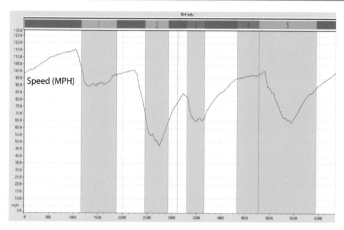

4.01. A speed trace showing that the fastest part of the track is just before Turn 1 at 115mph.

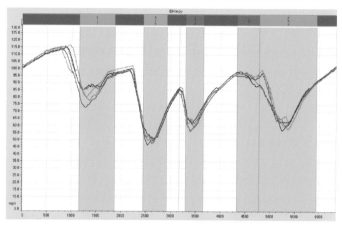

4.03. Overlaying speed traces from several laps gives a quick insight to the problem areas. Turns 1 and 5 need further thought.

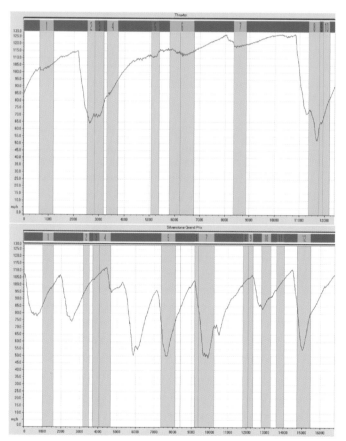

4.02. Speed traces from very different circuits The upper one is a high speed track with two major braking zones; the other much more technical, with nine separate turns.

are merely ways of getting to the objective. Speed is why we go racing, and is the measure of how effective we are.

It has a very characteristic shape and is shown in Chart 4.01. Not surprisingly the trace climbs and dips. At the risk of stating the obvious, the dips represent the corners and the climbs represent the straights. Viewed overall, it is your first indication of what sort of course you're running on. The chart here shows a short circuit with five turns: Turn 1 is taken at high speed, Turn 2 much slower, and Turns 3 and 5 are in the midrange. These corners all involve significant braking but Turn 4 is taken pretty much flat out, although judging by the shape of the speed trace, it gives the driver something to think about.

If we can look at the speed trace before going to a circuit it will tell us what to expect. The top half of Chart 4.02 shows a high speed circuit where aerodynamics is all important, and the lower half, one presenting a very different set of challenges.

Since speed is the end product, a very good starting point in analysing any session is to overlay several speed traces to get a feel for how consistent the car and driver are, and to identify problem areas. Chart 4.03 superimposes speed traces from five laps, and shows quite clearly that Turns 2, 3 and 4 are pretty much under control, but Turns 1 and 5 are causing difficulty. This gives us a useful insight into where to look for faster lap times, and sets the priorities in sorting the car.

ENGINE RPM

There are all sorts of reasons why we should be interested in engine rpm. Has the engine been over-revved? Is peak rpm being reached by the end of the fastest straight? Is the rpm too low (and, therefore, out of the power band) in any of the turns? When the driver shifts up (the rpm drops and then climbs again), has the engine dropped off the power curve?

In Chart 4.04 we can see the rise in rpm as speed increases along the straight, the fall under braking for Turn 1, and the jagged trace in the early part of the turn. This unsettled

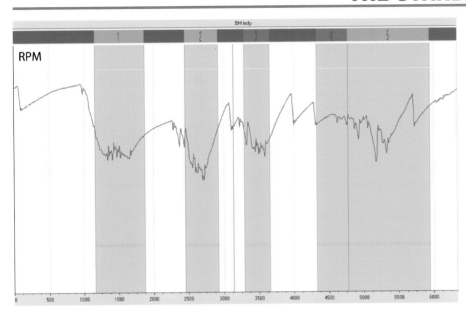

4.04. The rpm trace. The sawtooth pattern between Turns 3 and 4 indicates going up through the box, and the reasons for the jagged traces in all the turns needs investigating.

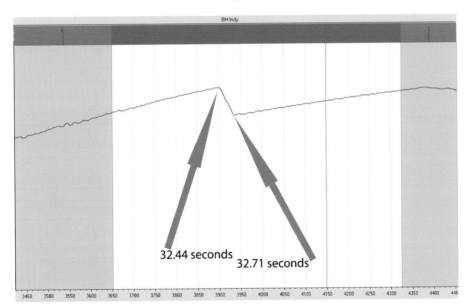

32.44 seconds
32.71 seconds

4.05. Zooming in on a gearshift. The logger shows the time elapsed in the lap, and this can be used to calculate the shift time.

trace is due partly to the bumpy track, and partly down to the car not coping well. Shifting up is indicated by the rpm rapidly increasing, falling steeply, and then climbing again. This can be seen best between Turns 3 and 4.

By measuring the time that elapses between the top and bottom of the drop due to shifting gear, we can get a feel for how well the driver can change gear.

One of the 'useful but not essential' software features in Chapter 1 was the ability to group data into one or more strips on a chart. It usually makes sense to plot rpm and speed on the same strip chart since they're related data, and you will often be interested in both at the same time: 'What speed was he doing here?', for example, and at what rpm?' Pulling these two together also saves screen space, which will get very crowded when you try work out what the data is telling you.

LATERAL G

Lateral g is the measure of cornering force. It is vital to most logging systems to enable a track map to be drawn if there is no GPS input, and to keep the GPS calculations honest when there is one. Track mapping is possible because, if you know the lateral acceleration (g force) and the speed, you can calculate the radius of the bend. It's a feature that comes in very handy when trying to understand the handling.

For the moment, the job should be to learn the shape of the lateral g trace so that you recognise it when it is comes up on the screen. It's not as distinctive as a throttle trace or an rpm trace, but it's essential if you want to know how the car is working. All the charts in this book use the convention that right-hand bends give positive g figures and left-handers are negative. This is contrary to strict engineering practice, but if the reader follows the trace, it curves to the right for right-hand turns.

The lateral g trace can give an insight into the line taken through the corner. The classical, constant radius racing line will be reflected by a lateral g trace that is symmetrical, building up and falling away in a steady manner. The more pragmatic (and faster) line that brakes deep, turns in late, takes a late apex and gets on the power early will show up as an asymmetric trace that climbs steeply and decays slowly.

STEERING

If you decide to use the right-hand-

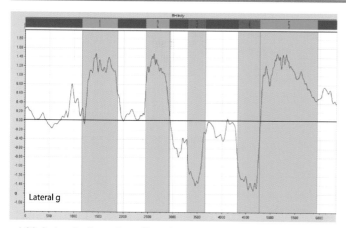

4.06. Lateral g trace from Brands Hatch Indy circuit. The shaded areas represent the turns, and right turns show a positive value (climb upwards on the graph).

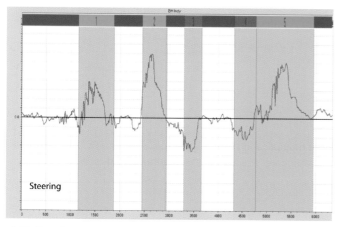

4.07. Steering trace taken from the same lap. Note that the shape is similar to the lateral g trace (it's the steering that causes the g!), but the jaggedness of the line indicates a nervous, oversteering car.

bends-are-positive idea, then it makes sense to use that for the steering as well. In that way, the steering and g forces move in the same direction on the chart, and you can understand more easily what is going on. Look at Charts 4.06 and 4.07 to get the idea.

THROTTLE

The throttle trace simply shows how much throttle is being applied and how the driver does it. It is an instantly recognisable one because it is full on for 60 to 80 per cent of the lap, and full off under braking, with a small transition stage as the driver feeds in the power. The scale is deliberately distorted on the trace shown in Chart 4.08 to make it more readable. The measurement is 0 to 100 per cent, but in this

case the axis is scaled as -20 to +500. This compresses the trace so that it can be shown on the same chart as other traces and makes full and zero throttle very obvious.

Bringing together this group of lateral g, steering and throttle on one screen helps highlight the relationships between them, and gives a first insight into how the car is handling.

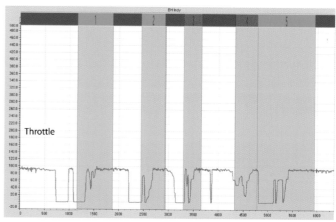

4.08. The throttle trace showing the lift-off before braking for the turns, a greedy driver who wants to be back on it as soon as possible, and who has to back out three times during Turn 1. Turn 2 is not much more restrained.

MATHS CHANNELS

The data that we can get from the logger is not just confined to the channels that we actually record. We can add to it by creating new channels mathematically. For example, if we do not have a sensor for Long g we could calculate it because the speed is known and so we can calculate the acceleration and express it in 'g'. If you followed the buyer's guide in Chapter 1, the software will provide the facilities to calculate this and a whole range of other things.

Maths channels are virtual channels – that is, they use existing data and re-present it in useful ways. Because they live off the physical signals that you collect, there need be no limit to the number that you can have. Some manufacturers disable the software and impose an artificial limit on the number of maths channels, but that's just so they can sell a professional version of the software at a premium. Maths channels can look daunting when you first bump into them or when someone shows you a favourite formula and you might well decide to add them to the list of jobs to avoid. But, in fact, being able to do calculations on your logged data is absolutely invaluable, and increases the power of the equipment tremendously.

A good example is 'miles per hour per thousand rpm' (mph/1000rpm) – a maths channel that shows the gearing, and

4.09. This is the sort of screen that you'll use to input the formula for a maths channel.

can indicate wheelspin or clutch slip, and tells you a lot about the driver's technique. Interpreting the trace is dealt with in Chapter 8, but our interest here is how to create a formula to provide the data.

There will be a route through the menus that gets you to a maths screen. Sorry, but the best way to find this will be to read the software manual. Whilst you have the manual in front of you take note that the available mathematical operators (+, -, *, /, integral, derivative, sin, cos, etc.) will also be listed and *almost* explained (that is, explained well enough for someone who knows enough computing and maths not to need much of an explanation anyway).

Anyone with any experience of using a spreadsheet will recognize the way in which a computer likes its numbers to be handled. It expects + and −, but also * to represent multiplication and / rather than ÷ for division. Numbers are raised to powers by the ^ symbol so seventeen squared (17^2) needs to be written as 17^2. The computer will do the calculations

from left to right except that certain operations are given precedence over others. It will always deal with calculations in brackets first; it will raise numbers to powers before multiplication and multiply before doing addition. Because of this we know that 2*3+4 will return an answer of 10 (2 times 3 equals 6, then plus 4 equals 10) rather than 14 (3 plus 4 equals 7, then multiply 7 by 2 equals 14).

To be sure to get the results that you intend, it's best to use brackets to enclose the parts that you want to go together. So, if you expect the answer 10 to the previous calculation you should type it as (2*3)+4. If 17 was the answer that you were looking for the expression would be 2* (3+4)

A typical screen will look something like that shown in picture 4.09:

Getting back to mph/1000rpm, the formula to calculate it is expressed as:

mph/1000rpm = (speed) ÷ (rpm ÷ 1000)

A few things need commenting on. First, you can invent a name that

you feel comfortable with and in this case mph/1000rpm is easy to say and conveys what we mean perfectly well. Since we are programming the computer to do a calculation for us, we have to observe its rules about names and syntax. So the mph/100rpm formula would need to be presented to the computer as

$$mph/1000rpm = \frac{(Speed\_1)}{(Engine/1000)}$$

The formula now uses the actual channel names of the variables, so here 'Speed_1' and 'Engine' are the names given to the wheel speed and rpm channels when the logger was installed. Finally, the brackets are there to make sure that the computer does the calculations in the required order. There are more brackets here than necessary, but it's better to be safe than sorry.

If you decided that rather than mph/1000, you would prefer to log the gear ratio, then the formula for the maths channel would be slightly more complex but still using the same basic ideas.

Gear ratio = (rpm/60)/ ((speed * 0.44704)/ wheel circumference)

To explain, the (rpm/60) calculates how many revolutions the engine does in one second. Putting ((speed * 0.44704)/ wheel circumference) in a containing pair of brackets tells the program to deal with two distinct halves to the calculation. The speed * 0.44704 converts mph to metres per second and, if the wheel circumference is measured in metres, then speed in units per second divided by circumference will give us the number of wheel revolutions that take place in one second. Dividing the engine revolutions per second by this number gives us the gearing ratio for the car.

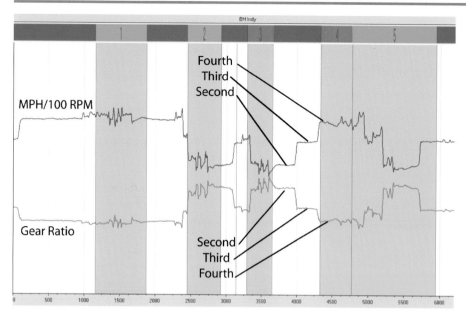

4.10. Mph/1000rpm and gearing are two simple maths channels that indicate the overall gearing of the car.

This formula calculates the gearing in terms of number of engine revolutions for each revolution of the driven wheel. It also gives insight into wheelspin, clutch slip, and driver technique. This gearing calculation results in an identical trace, albeit inverted, to the mph/1000. They are shown in Figure 4.10. In an ideal world they would both produce crisp square lines that stepped up or down each time the driver shifted gear. The fact that they don't tells us about the way the car and driver are performing.

LOGICAL AND OTHER OPERATORS

A typical program will provide a range of mathematical operators above and beyond the ones in the previous paragraph. Most systems seem to have a pretty full complement of trig functions (sin and arcsin, cos and arccos, tan and arctan), the ability to pick out minimum median and maximum figures and to calculate derivatives and integrals. In addition there are usually logical functions such as 'if' or 'greater than'.

One bit of bad news is that even if you avoided Calculus at school, you're going to need a bit of it for logging purposes. The first important idea is that you can use differentiation to measure a rate of change. The rate of change of speed is acceleration, and we're going to want to know that at some point. Fortunately, the logger will do the calculations if we tell it what we want to know. If we want to know acceleration of the car all we need to do is to differentiate the speed. To do this we just need to use a function called the derivative function, as follows:

Acceleration = deriv (Speed_1)

This will work out the change in the value of speed per unit of time. Here, the derivative function is named 'deriv', and it needs what computer people call an 'argument'. The 'deriv' function needs to be told what it is to be applied to, and you do this by enclosing things in brackets. If the speed is measured in metres per second, the derivative will be in metres per second per second. If the speed is in feet per second, the value would be expressed in feet per second per second. We know that the acceleration due to gravity is 9.81 metres per second per second, or 32.2 feet per second per second, so converting the speed into metres per second and dividing it by 9.81 would give us an answer in g. Converting the speed into feet per second and dividing by 32.2 would give the same number. So, if the speed channel is measured in miles per hour, the following two equations would calculate acceleration and express the answer in g:

deriv(Speed_1*0.44704) / 9.806

or

deriv(Speed_1*1.467) / 32.2

Here the speed in mph is converted to metres per second by multiplying by 0.44704 and then converting to g by dividing by 9.806m/sec (the acceleration due to gravity). An equivalent process would be to convert speed into feet per second by multiplying it by 1.467 and dividing by 32.2ft/sec. Readers who measure speed in kilometers per hour will need to divide this speed by 3.6 to convert km/h to m/sec, and use the 9.81 for acceleration.

By doing this calculation we have created an acceleration channel that we will call Long g for the rest of the book. An interesting use of the derivative function is to assess driving style. We could calculate the rate at which the driver was getting into the throttle by simply using:

Throttle speed = deriv (throttle)

The 'deriv' gives us the rate of change of the throttle position. A low number (below 75 per cent per second) would indicate a timid driver, whereas a high number (in excess of 150 per cent per second) might be just too aggressive. Naturally, the boundary between assertive and downright clumsy will vary according to the balance between power and grip, but it is another area where the data can help understand what is going on. This rate of application can also apply to brakes (see Chapter 9) and to steering (Chapter 7).

After differentiation (the derivative) the other bit of calculus that you should come to terms with is the idea of integration. This adds up all the individual bits of data to give us the integral. The program will do this for us. It can be applied to the speed channel to calculate distances, for example. If we wanted to know how many miles had been covered the maths channel would look something like this:

Distance covered = integral (speed)

The software would calculate the integral of the speed by looking at the speed during a time slice (dependent on the sampling rate), work out how far the car has travelled in that time, and add up all these small distances together into a total. So, if you were sampling at 20Hz, the calculation would be done every twentieth of a second.

The programs calculate the integral of a number, either for the lap or for the whole run. So, if we wanted to know how much time was spent per lap under braking, we would use the lap integral. If we wanted a running record for the whole session we would use the run integral.

This idea is used widely by professional teams to 'life' components, so they would sum the time that, say, first gear was being used and keep a track of that information to ensure that the first gear does not exceed its safe working life. Recorded at the end of each day's running, we would know how many miles the car had covered.

Some software distinguishes between the integral for the lap and for the whole session. So the lap integral would be useful for calculating the amount of time spent on the brakes for each lap, but the session integral would allow you to calculate the number of hours that you ran the engine during the session.

Looking now at the logic functions, they will allow us some degree of choice in the calculations. Spreadsheet users will be familiar with the 'if' function where the syntax is:

If (condition, then, else)

This simply means that the program can test to see if a particular condition is being met and, if it is, then something happens, or else, if not, then a different choice is made. Some systems call this the 'choose' function, but it works in the same way as the 'if' function.

For example, this could be useful in calculating brake balance because there's always a little bit of 'noise' in the system. The pedal rattles about and adds a very small amount of pressure to each circuit. If there were, for example 0.05 bar in the front circuit and 0.07 bar in the rear, and a calculation of brake balance would show a balance of 41 per cent to the front when in fact there was no real braking going on and we should have no interest in the brake balance. So the three steps to the 'if' are

1 If the value is less than the chosen limit (that is the condition)
2 then do nothing
3 or else include it in a calculation. This would be shown as

If(brake front <0.5, 0,brake balance calculation)

This would give us a maths channel for brake balance. In fact, because this was a logical test included in the

calculation of the brake balance, the formula will actually look more like this If(brake front<0.5, 0,(brake front/(brake front + brake rear))*100)

This creates a maths channel called brake balance using the following stages:

1 If the pressure in the front brake circuit is less than 0.5 bar
2 Then show a value of 0 for the brake balance maths channel
3 Or else calculate the brake balance as front brake ÷ the sum of front and rear brakes and multiplied by 100 to create a percentage figure.

Again, anyone with a familiarity with using spreadsheet functions will feel reasonably comfortable with this sort of logic. For newcomers, the 'condition' and the 'then' and 'else' should, by now, be beginning to make sense, but the formula to calculate the brake balance is liberally sprinkled with brackets. The black pair encloses the 'argument' or parameters of the function; the red set separates the fractional brake balance from the fact that it is to be multiplied by 100 (to give the answer as a percentage) and the green set makes sure that the front and rear are added together before anything else happens. Otherwise the computer, using its inbuilt order of precedence, would multiply the rear brake pressure by 100, divide the front pressure by the rear pressure and add the two results together. Not what was wanted at all. It's always good practice to spray the brackets around in formulae and functions; it helps you clarify your thinking and minimises the risk of error.

Another way to do this is when the sensors are being calibrated. You can usually decide on a limit (or threshold) below which data will not be logged. The drawback with that method is that you are deliberately letting go of data before it ever reaches the analysis stage, but with the software based approach you can see what the data is doing before you decide whether or not to lose it.

CLEANING THE DATA

Data, whether it comes straight off the logger or out of a maths channel can often be very messy with 'noise' in the trace that makes it difficult to understand. Some of this is because the sampling frequency was too high and some is simply a matter of the behaviour of the equations.

The data can be made more legible by adjusting the sampling rate or by mathematical smoothing over a time period or a number of samples. These two have different effects on the way that the data looks when it's displayed. It's very tempting, if you have the technology and the memory resources to gather data at the highest frequency possible. So, for example, lateral g might be sampled at 100hz (100 readings per second) in the hope that what you'll get is a very detailed record of lateral g. You will also get a very noisy signal, and the extra data just makes the trace look more messy. If you look back to Chapter 1, Chart 1.02 shows the difference between logging at a high and low rate.

A related consideration is the extent to which the data should be smoothed or filtered. The software will offer you choices – either in terms of the time over which the smoothing operation should take place, or the number of samples. There is a range mathematical models that can be used to smooth off the data, and it seems that there are differences in results between different analysis programs. Some users prefer to use heavy amounts of filtering in order to damp out even small disturbances that are caused by random events, such as bumps on the track surface, cross winds, etc.

The simplest form of filtering is where the software creates a moving average that includes the data point in question and a given number of points either side of it. You'll be asked to specify a value for the filter to choose the degree of smoothing, and it's a sensible precaution to choose an odd number. This is because we don't know precisely how the program will carry out the smoothing operation. Choosing an odd number means that if the software puts our data point in the middle of the range, we are sure that there can be an equal number each side of it. So choose 5 and this might mean that there is an average of five values taken, two on each side of the central data point. Choose 4 and it might mean that there are four points chosen, one before our point, and two after. If that happens, there is a time shift, because the average is moved slightly forward in time. It might equally mean that there are four data points each side of the central point, but since we don't know for certain, it's good sense to treat with care and investigate exactly what effect the program has on your data.

Pass filters are also fairly common and simply put limits to the data. A low pass filter allows through only the low values so that you choose a high threshold above which all the values will be ignored. The high-pass does the opposite and the band pass filters the values in the middle range. Sadly, the manuals are not always crystal clear as to what exactly is going on when you apply the filter and you also need to be wary of is whether once you have carried out the filtering operation you have actually replaced the original data with new, filtered data. Until you are sure what is happening to your data it is best to create a new channel with its own distinct name.

Chart 4.12 shows lateral g in the

AIM software as a raw signal logged at 10hz, left unfiltered and shown in red. This red trace is largely hidden by a blue trace which is the same data filtered at the maximum level available in the software. For most of the time the two traces are identical and any difference only shows up at the extremes where the values of the peaks and troughs are reduced (in this instance) by about 3 per cent.

If you have a mathematician on the team and your software supports the more sophisticated filtering techniques, such as the Chebyshev method, or windowing techniques, such as Hamming or Bartlett, you should take advice as to which is appropriate for your purposes. If not, it is safer to operate on the 'suck it and see principle'. Look at your data unfiltered and then apply varying levels and types of filtering to it and evaluate what the effect is. Make your own choice as to the right level for the job that you're doing, but make sure you know what is happening to the data once it leaves its raw state and is re-presented to you modified by a filter.

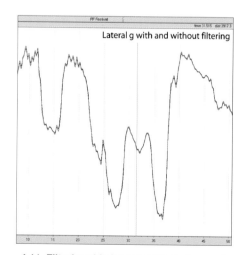

4.11. Filtering this lateral g trace does not make much difference to its meaning.

Chapter 5
Displaying the data

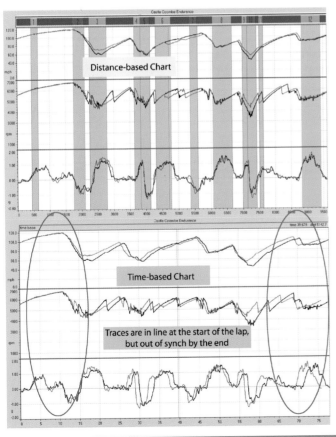

5.01. Two choices for the X axis.

There are many other ways of presenting the data in addition to the traditional strip chart, and this chapter looks at a few of them. There are also some suggestions as to how various types of charts and graphics can be pulled together on to one screen, to give a range of ways to view the data. These pre-defined screens speed up the analysis process on race and test days.

The strip charts shown in the previous chapter were all plotted against distance, and this is conventional. It allows one lap to be superimposed on another for comparison. You are always offered the option to plot against time and, if you do this, one lap will lag behind the other and the traces will get out of sync. Chart 5.01 shows the effect. Generally, there's little reason to use time as the base except for special investigations.

TRACK MAPS

Track maps are a good way to start any analysis because they put the data into context. Naturally, GPS-based systems do very well here by creating a very credible map with little or no effort. With a bit of experience, your feeling for what part of the track the data relates to will increase, and less reliance will be placed on the map, but it's still nice to have a good map of the circuit on screen and on paper for note taking.

Systems that use speed and

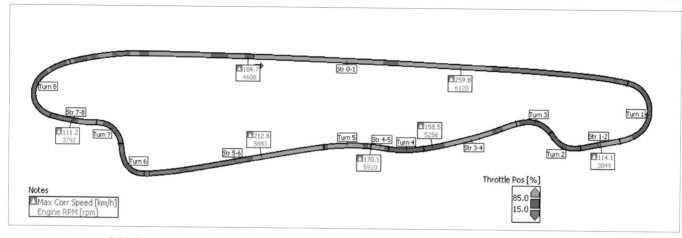

5.02. A track map shows a variety of data with immediate impact. Not all systems are this good.

lateral g data to draw the map will need some attention to generate a decent representation of the track. Some systems are better than others and can do a reasonable job with little manual interference. Some will come up with a first attempt that looks little or nothing like the circuit, and you'll have to add in various fudge factors to overcome nonsensical output, such as the track crossing back on itself and the wrong general shape of the circuit. It's worthwhile taking the trouble to create a decent track map (using one of the early warm up or sighting laps) because, once done, it's on file for future use. There is a reassurance in a map that reflects what you know to be the case.

Be warned, though, that if you have difficulty in getting a reasonable track map out of your data and other people with the same software do not, then you should think about the calibration of the sensors. The less accurate your speed and lateral g figures, the odder the map will look, and the more difficult it'll be to get it into shape. This is not a critical indicator, just a nudge in the direction of checking your settings. When you're satisfied with the quality and accuracy of the map, it can be printed out as a basis for note taking and record keeping.

Modern systems can do much more

than this. They can shade the map to reflect the data so that you can get all sorts of extra insights. Some can add notes of data readings at points on the circuit. Figure 5.02 is shaded to show the throttle position, with green being greater than 85 per cent, red less than 15 per cent, and blue the bits between. It gives an instant reminder to the driver of braking points (start of the red areas), it indicates gear changes (short blue patches) and areas of full and balanced throttle application. This software program also allows other data to be indicated, in this case, speed and rpm which are added as notes.

TIME SLIP
One of the clever ways that the logger has of showing us where time is made and lost on the track is known as the 'time slip' line. It's a distance-based strip chart that shows a reference lap as a baseline. This is usually the fastest lap, but some software gives you a bit of choice. The chart then has another line that represents the time difference on the lap that you're comparing. If the comparative lap is above the baseline, the time difference is positive (you're losing time) and so it's slower than the reference lap. If it is below, then the comparative lap is quicker at that point.

For interpretation purposes, the simple rule of thumb is that if the time slip line is heading upwards, time is being lost and it's bad news. When the line heads down, then the slower lap was making time on the faster one and this is good news. It shows instantly and graphically where time is being made and lost. It is usually possible to compare several laps at once.

Figure 5.03 shows how this works in practice. The top trace shows speed, the lower trace time slip and the fragment shows two corners linked by a short straight. The speed trace on its own tells the story, but the time slip shouts it out loud and clear. The blue trace is better because, although the driver brakes about 40 metres earlier for Turn 1, she is able to carry more speed through the corner. This gives more speed on the short straight connecting the two bends and a higher speed before braking for Turn 2. Earlier braking and a slightly slower passage through Turn 2 allow her to get back on to the power 0.2sec earlier and to carry that speed into the early part of the straight. That early application of power on the blue lap is not compensated for by the higher cornering speed through Turn 2 of the red lap.

The time slip trace shows time

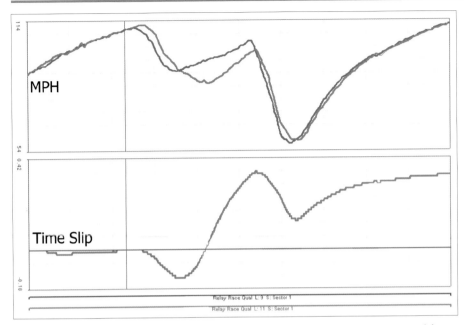

5.03. The time slip trace shows where time is being saved (trace heads downwards) or lost (trace heads upwards).

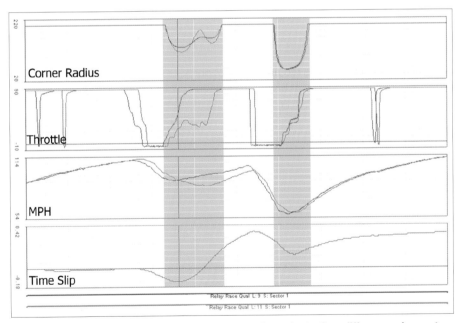

5.04. The time slip trace shown with other data to investigate time differences in greater detail.

saving of up to 0.13sec. That's where it all starts to go wrong. Braking later meant that the driver felt it prudent to be on the brakes for longer and, in fact, she shed more speed (down to 85mph as opposed to 90 for the blue lap). Entering the corner too quickly also meant that the driver couldn't get back on the throttle as early, so speed didn't build as quickly on the red lap as it did on the blue.

By now the time slip line has climbed past zero (undoing the immediate benefits of the late braking) and peaks at about 0.35sec. It was probably knowledge of this error that persuaded the driver to brake 0.6sec later for turn 2 and carry more speed through the turn, and this is shown by the time slip graph heading downwards as time is clawed back. The unfortunate effect of this higher speed is an inability to get back on the power until things have calmed down a bit, and she gets back on the throttle about 1.2sec later and the time slip starts to climb again.

We could back up these interpretations with additional evidence. The Chart 5.04 is the same data, supplemented with corner radius and throttle position. The corner radius shows that she adjusted the line that she took through Turn 1. It also shows later braking and application of the throttle in Turn 2.

The software generally tries to be helpful by auto-scaling the time slip line. So, if the trace is a standard size, the time saved is expanded to fill the space available, and an identical line height might represent 0.4sec on one screen and 2 seconds on another. If one lap is only half a tenth slower than another, you might find yourself analysing slopes and peaks that only amount to a few hundredths of a second and really not worth the effort. This is irritating until you train yourself to look for it.

gained and lost even more dramatically. For the red lap, the drive out of the previous corner was better and the speed 1 to 2mph better. This is reflected in a time slip graph that shows the red trace just dropping below the zero line and the saving of a couple of hundredths of a second. Late braking on the red lap is shown as the time slip line moving below the zero line and results in a

			1		2		3		4	5	
Absolute split times											
200 Oct 18 BH Indy S1	run 2 lap 2	7.689	6.312	4.715	6.470	3.387	3.790	5.763	3.627	11.548	2.956	00.56.258
200 Oct 18 BH Indy S1	run 2 lap 3	7.591	6.266	4.855	7.831	3.852	4.181	5.796	3.957	11.518	2.967	00.58.813
200 Oct 18 BH Indy S1	run 2 lap 4	7.514	6.343	4.427	5.959	3.270	3.757	5.540	3.439	10.887	2.967	00.54.103
200 Oct 18 BH Indy S1	run 2 lap 5	7.471	5.688	4.292	5.844	3.282	3.573	5.431	3.201	10.761	2.957	00.52.501
200 Oct 18 BH Indy S1	run 2 lap 6	7.406	5.538	4.203	5.838	3.297	3.647	5.455	3.173	10.376	2.881	00.51.814
200 Oct 18 BH Indy S1	run 2 lap 7	7.307	6.076	4.370	5.752	3.262	3.990	5.785	3.234	10.424	2.899	00.53.100
200 Oct 18 BH Indy S1 (be	run 2 lap 9	7.321	5.277	4.162	5.836	3.298	3.593	5.492	3.161	10.186	2.922	0.51.247
200 Oct 18 BH Indy S1	run 2 lap 10	7.301	5.460	4.146	5.809	3.301	3.990	5.737	3.221	10.207	2.894	00.52.065
200 Oct 18 BH Indy S1	run 3 lap 2	7.425	5.775	4.377	6.268	3.416	4.078	5.566	3.285	10.426	2.905	00.53.522
200 Oct 18 BH Indy S1	run 3 lap 3	7.417	6.132	4.418	5.839	3.229	3.767	5.371	3.208	10.421	2.841	00.52.642
200 Oct 18 BH Indy S1	run 3 lap 4	7.362	5.842	4.193	5.835	3.215	3.757	5.361	3.826	10.664	3.099	00.53.154
minimum value		7.301	5.277	4.146 [s]	5.752 [s]	3.215	3.573	5.361	3.161	10.186	2.841	
maximum value		7.689	6.343	4.855 [s]	7.831 [s]	3.852	4.181	5.796	3.957	11.548	3.099	
average value		7.436	5.872	4.367 [s]	6.106 [s]	3.344	3.816	5.562	3.382	10.645	2.936	
std deviation		0.117	0.353	0.220	0.583	0.170	0.198	0.165	0.273	0.465	0.064	
Theoretical best lap											
2007 Oct 18 BH Indy S1	best	7.301	5.277	4.146	5.752	3.215	3.573	5.361	3.161	10.186	2.841	0.50.813

5.05. A lap and split report showing the fastest lap (51.247), the fastest rolling lap (51.227), and the fastest possible lap (50.813).

SECTOR AND SPLIT TIMES

The other approach to the times is by using the logger to give you a detailed printout of lap times and sector times. Split times attract a lot of attention because everyone, especially the driver, wants to see what the lap times were and what the time would be for the fantasy lap made up of the quickest bits from all the laps. Whilst this is the fastest possible lap, it's more realistic to be interested in the fastest rolling lap and the spread of lap times. It's a good day when every lap is faster than its predecessor, but this doesn't happen very often.

The fastest possible lap and the fastest rolling lap perhaps need some explanation. If the lap is split into sectors, then it is possible to add all the best sector times together to get a fastest possible lap. Every driver loves the thought that they are actually a second or two faster than the official timing shows, but that may not be realistic. The big problem is that the driver may absolutely fly through (say) Sector 7 and arrive at the beginning of

Sector 8 with the car on the wrong side of the track and in no position to turn in where required. This leads on to a slow time for the Sector 8 which is discarded in the adding up of the fastest ever lap. To put together a really good lap, Sectors 7 and 8 need to be considered together to be realistic. On the final lap recorded in Chart 5.05, the straight between Turns 3 and 4 was 0.13 sec faster than on any other lap. This obviously compromised Turn 4 which, on that lap was 0.66sec slower. This is not such a problem if you use fairly long sectors and split the lap into, say, three or four parts. Then the transition between sectors becomes less significant.

It's much more realistic to pay attention to the fastest rolling lap. This takes the fastest sequence of sectors irrespective of where they start. So it might be that Lap 4 was pretty messy in the first part, but got better towards the end. If Lap 5 started well, the combination of the back of Lap 4 and the start of Lap 5 were combined, the two represent what the car and driver are actually capable of even though it does

not coincide with the official timing and will not earn you the lap record.

What is important is that the times show some degree of consistency. If not, then there's something wrong with either the car or the driver. If there's inconsistency lap-to-lap, it's a simple matter to create a new chart that shows speed, and to overlay as many laps as the software allows. This shows quickly where the problems lie and can direct your attention to further data. For instance, if there's inconsistency under braking, then you know to start looking at this sort of data. Chapter 4 showed a strip chart (5.03) that overlaid several laps in order to find inconsistencies. The detailed split report shown here quantifies the extent of the inconsistency.

CHANNEL REPORTS

The logger will also prepare channel reports (which can usually also be pre-configured) that can show key statistics for any channel that we might be interested in. They provide information not just for assessing how things are

Lap	avg	avg	max	max	min	max
	Speed_1	Throttle	Front Brake	Rear Brake	Corner Radius	Brakes used for
run 1 - lap 3 03.12.343	37.1	19.2	19.6	36.5	-155.00	12.50
run 1 - lap 4 01.31.826	77.6	52.4	12.5	30.7	-155.00	11.10
run 1 - lap 5 02.01.441	58.2	21.3	11.5	24.0	-155.00	8.00
run 1 - lap 6 02.03.517	57.4	24.4	10.7	20.1	-155.00	13.40
run 1 - lap 7 01.20.039	87.6	58.1	16.0	31.7	-155.00	10.90
run 1 - lap 8 01.19.278	88.4	58.8	14.4	33.3	-155.00	8.90
run 1 - lap 9 01.19.486	88.0	61.7	17.4	40.3	-155.00	9.80
run 1 - lap 10 01.20.154	87.3	63.1	20.5	41.3	-155.00	10.50
run 1 - lap 11 01.19.351	88.2	60.2	12.7	32.6	-155.00	9.00
run 1 - lap 12 01.18.756	88.9	66.6	15.1	33.5	-155.00	9.30
run 1 - lap 13 01.18.276	89.5	64.5	13.2	32.4	-155.00	8.20
run 1 - lap 14 01.17.400	**90.5**	**66.9**	**15.1**	**34.0**	**-155.00**	**8.30**
run 1 - lap 15 01.18.537	89.1	64.7	18.4	35.7	-155.00	9.10

5.06. A typical channel report, this one showing braking data.

behaving on that day, but they are also part of the set of records that all competitors should keep. Engine health is an obvious choice here and we would be interested in maximum, minimum and average values for RPM, pressures and temperatures. This would show at a glance whether the engine is operating within the correct limits. You can get a quick assessment of the handling by providing key stats for the various g forces and top speeds. The same goes for just about any area of interest, such as braking, driver inputs, and so on.

Inevitably, the software differs between makes, and some is more sophisticated than others. For example, some software will not only show the min, max and average for a channel, but it will also split it over the segments of a lap that were used in the lap time reports. It's possible with some software to identify percentiles, so you might choose to show the 95th percentile (the figure within which 95 per cent of the values fall) so that you eliminate the most extreme readings. Others show the standard deviation so that you can get a measure of how dispersed (scattered) the data is.

HISTOGRAMS (BAR CHARTS)

Histograms are charts that show the percentage of a particular variable occurring within specified bands (the software might call them 'bins'). They can be applied to any variable, but are useful for rpm, speed and throttle. It's very worthwhile creating a standard screen that allows you to call these up at a single mouse click, and one such is shown in Chart 5.07. The significant ones are:

• Engine rpm
• Speed
• Throttle position

Rpm can be useful because it shows the amount of time that the engine spends in each part of its range, and the histogram provides a quick way of checking that the engine is not being asked to operate out of the power curve or above its comfortable limit. It gives some insight into whether the gearing is right, so is one of the things to take into account when making gearing decisions.

Chart 5.08 shows the result of a gearing change. The black histogram represents 'before' and the green one

'after'. As can be seen, the result is that after the ratio change, there was no need to hang on to the gears beyond 7000rpm as there had been earlier. The car is able to spend more time in the productive 6 to 7000rpm band and not run the risk of over-revving.

Speed tells us something about the nature of the circuit by displaying the amount of time spent in each speed range. Chart 5.09 shows the different speed ranges of two circuits. As such, it can point us in the direction of things to concentrate on in setting up the car for a competition. If a lot of time is spent in the lower speed ranges, then there are gearing issues, but it also tells us that perhaps traction and agility are the important things to aim for in the set up. Comparing histograms from different sessions will give us an insight into the effects that changes are having.

A throttle histogram has a distinctive shape, with the majority of time spent with the throttle wide open and a significant minority with it firmly shut. Not surprisingly, the fastest lap almost invariably comes from the one with most time spent with the throttle wide open – a fine example of data

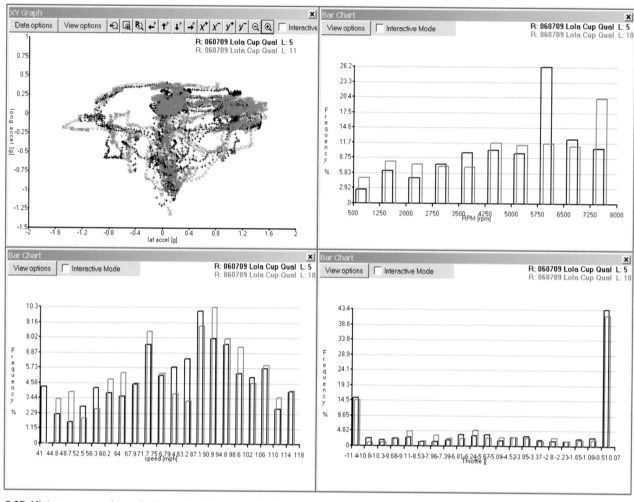

5.07. Histograms can be pulled together into a standard screen. Here, rpm, speed and throttle are shown with a friction circle to complete the picture.

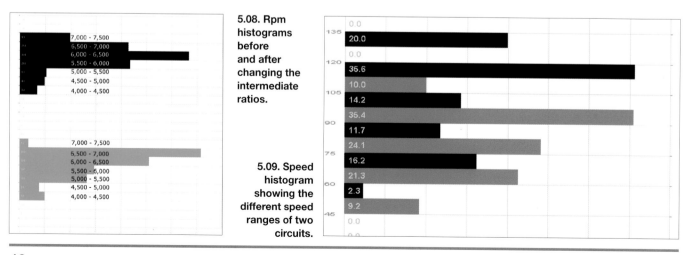

5.08. Rpm histograms before and after changing the intermediate ratios.

5.09. Speed histogram showing the different speed ranges of two circuits.

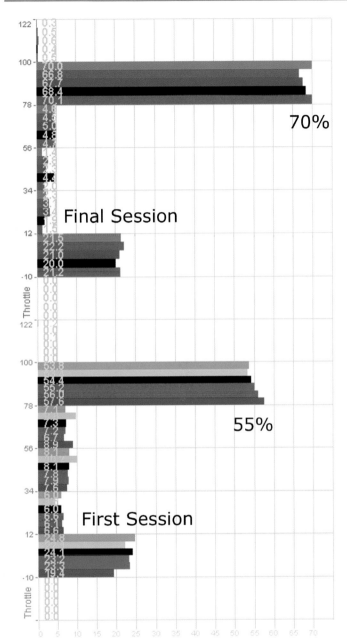

5.10. Before and after: the improvement made over a day's testing shows in the throttle histogram, with the upper chart displaying the data from the last session of the day.

logging telling us the blindingly obvious. Time spent between the two (shut and wide open) is a good indicator of how difficult the car and circuit are, and time spent improving the car will result in a smaller area in the middle of the histogram.

Chart 5.10 shows histograms taken from a number of laps in the opening and closing sessions of a day's testing. The

opening session is the lower of the two histograms and shows a significant amount of time spent in the midranges. The time in the midrange has reduced significantly by the end of the day to the benefit of time spent at full throttle. The day in question was very busy, and the track lost grip as the day progressed. In spite of this, the driver was happy that progress had been made, even though it didn't show up convincingly in the lap times. The throttle histogram indicated that he had more confidence in the car in the final session.

XY GRAPHS

XY graphs are a way of comparing one variable with another. Instead of having time or distance on the X (bottom) axis, you can choose whatever variable suits you. They can often

	Chart	X Axis	Y Axis
Friction circle	11.08	Lateral g	Long g
Gear chart	8.02	Speed	rpm
Aero effects	7.02	Lateral g	Speed
Handling	11.11	Steering	Lateral g
Brake balance	9.09	Front brake pressure	Rear brake pressure
Oil surge	9.05	Lateral or long g	Oil pressure

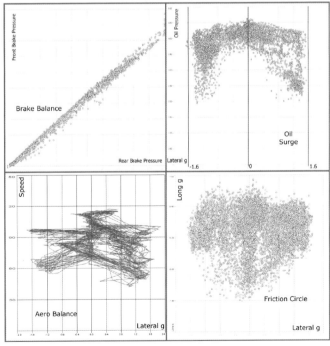

5.11. Four XY charts showing some typical applications.

present you with information in a surprising way, and they're a useful way of simply poking about in the data to see what pops out. Some of the standard XY charts are shown in the accompanying table.

STANDARD SCREEN LAYOUTS

One useful way to streamline the data handling is to create a standard set of screen or chart layouts that you know will be useful. This is an area where there are some differences between software packages. Some will allow you to create quite complex screens with a mixture of different types of display all on the same view. Others only provide limited facilities for creating standardised designs. Some manufacturers go to great lengths to provide standard layouts that you can adapt. Others leave you on your own.

If you can pre-fabricate screen layouts, an evening's work will equip you with a set of screens that will fit your needs precisely and speed up your work at the track. It will be time well spent. Each screen can bring together all the data that you need regarding a particular area of interest. For example, 5.12 shows Stack's pre-defined screen setup relating to gearing. The strip chart shows the detail for close investigation, the XY chart of rpm vs speed shows the relationship between the gears, and the data table shows relevant values at any point on the map.

The next chapter discusses ways to structure your approach to the data so that you can use it most effectively. It suggests a sequence of dealing with the data that reflects the realities of running a racing car. This involves a quick check to see that everything is working properly and then searching systematically for evidence of problems. You will no doubt evolve your own way

of doing things, but in the meantime here are a collection of standard charts to start with:

- Engine health
- Gears and speed
- Driver inputs
- G forces
- Brakes

Engine health is simply a matter of seeing that rpm, temperatures and pressures stay within reasonable bounds. Generally, they are pretty unexciting to look at. Temperatures can be expected to climb slightly throughout a session, and possibly pressures may fall. Battery voltage will fall away slowly if there's no charging system, but be more agitated if there's a generator. The charging system will cause the voltage trace to step by a small amount every time that the alternator kicks in. Unless there's a known issue with the engine, it deserves no more than a quick look through. This is one case where plotting against time rather than distance can make the picture clearer. Chart 5.13 shows the warm up on the left and some time spent in the pits.

Gears and speeds come next. Chart 5.14 shows speed, rpm and mph/1000rpm. An obvious first check is whether the expected speed was reached at the end of the longest straight. It's an opportunity also to look at the engine rpm just before braking. You should expect to see a rpm beyond the power peak and approaching the safe rpm limit. You can also do quick a review of the ratios to see what gear is used where, whether there are any corners where the engine is being operated outside its comfort zone, and whether there are problems with having to change up or down unnecessarily.

The data presented here shows the engine just about maxing out at the end of the longest straight and not dropping

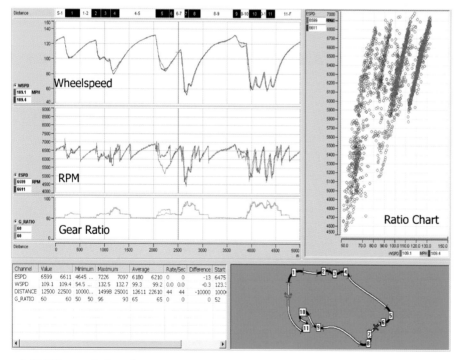

5.12. This pre-designed screen brings together all the relevant data for making gearing decisions.

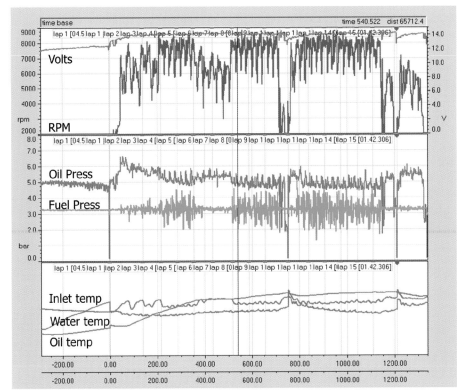

5.13. This engine health chart shows all the laps of a test session. Pressures and temperatures stay within limits through the period. The rpm trace shows when the car was not running.

the handling is being influenced by the driver. There are messages available to us in these charts and all of the matters raised are dealt with in greater detail in later chapters.

It is also worth looking at the overall level of performance by looking at the g forces that are generated. These are dealt with in more detail elsewhere in the book, but the data shown in Chart 5.16 is used as a quick check that the car is achieving the levels of grip that it ought to. You need a benchmark to compare the performance with, and this will be based on previous performances with this car. Not all tracks have the same amount of grip, and not all days are the same at the same track, so it's useful here to be able to share data with someone else to reassure yourself that the figures are OK. The top trace represents cornering force, and the middle one is acceleration and braking. The bottom trace is combined g forces, and is a measure of how well the driver is using the tyres. If your car has significant downforce, it's worth including a speed trace on this chart so that the aerodynamic effects can be assessed.

out of the power band in any corner. Because this screen uses mph/1000rpm rather than a formal gear counter, the extent to which that trace varies from a set of clean angular steps is an indicator of driver style and car traction. The traction problems and poor technique that show up here are dealt with more fully in Chapters 8 and 11. Here we can see that the driver is hanging on to second gear and reaching high rpm along the third straight, and on the fourth is making two upshifts. This is evidence that we need to give some consideration to the intermediate gear ratios.

Now we can turn to analysing the performance, and the driver inputs (Chart 5.15) are a good starting point. The mph/1000rpm shows what gear is being used at various parts of the circuit. The required and actual steering is a broad measure of how well the car is handling,

and tying this in with the throttle position gives us a clue as to the extent to which

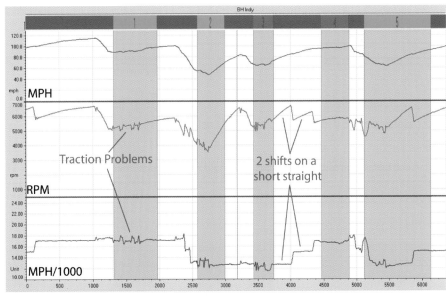

5.14. The gears and speeds chart highlights some issues that need attention.

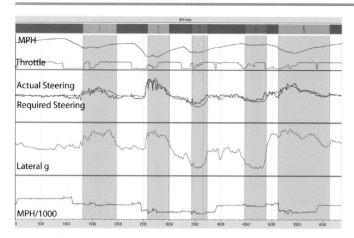

5.15. Driver inputs. This view shows what the driver is doing. Required steering (blue) is a maths channel that shows the steering angle needed for the path being followed by the car. Actual steering is recorded directly.

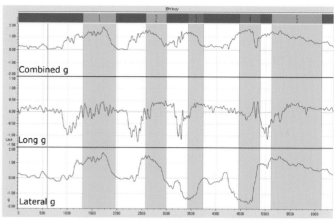

5.16. Looking at the g forces generated gives a feel for the overall levels of performance.

Combined g is simply the sum of lateral and longitudinal g and is calculated as:

Combined g = $\sqrt{(\text{lateral } g^2) + (\text{longitudinal } g^2)}$

It's a measure of the extent to which the driver is using the tyres to the maximum, because it should show a series of 'humps' that rise when the braking starts and stay at a high value during the transition into the corner. A dip at this point indicates that the driver is not making full use of the tyres. The same applies to the exit when the reduction of cornering forces should blend into forward acceleration.

The final screen is one that shows the braking performance. This is not going to make a huge difference to lap times and the benchmark should be previous performance to make sure that everything is operating well. The chart in 5.15 shows only the basic data but it would be possible to show other information as well, such as the amount of time spent

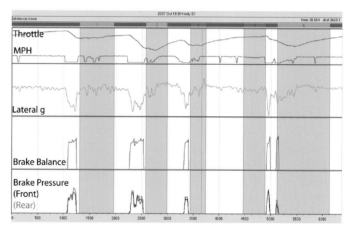

5.17. Braking performance can be reviewed by pulling together all the relevant data onto one screen.

coasting. Chapter 11 looks at how to identify the time spent coasting (using neither the throttle nor the brakes) and the overall time spent on the brakes.

Chapter 6
Understanding starts here

The aim of this chapter is to show you some quick ways into the data. A typical run can yield half a million data points that you need to make sense of as soon as possible, so here are some strategies that work. In reality, you've only a limited amount of time in which to turn the car round so you need to squeeze as much information as possible out of the data as quickly as possible. Over time, you'll develop methods that work for you. Until then, this chapter presents a way of getting the first messages out of the data so that you know what to do next.

Every time that the car is run the data should be downloaded and the following stages gone through.

• A quick check to see that everything is functioning properly
• Looking for evidence of problems through multiple overlays
• Review from the driver's perspective
• A longer, more leisurely search of the data (on the day, if time is available, or later, back at base if not)

The first three of these stages are dealt with in this chapter, but the driver's perspective is looked at more fully in Chapter 11, and the more considered and leisurely run through the data is dealt with in the 'After the Event' section in Chapter 10.

The whole team should be interested in what the data has to show. It's also best to give someone permanent responsibility for downloading, and preferably this should not be the driver (they are usually full of adrenalin when they get out of the car and might well break something). The best bet is to send the driver off on a five minute paddock walkabout while the first stages of reading the data are carried out.

Everyone will want a piece of the data, so the absence of one team member (the driver) at this point is not a problem. The mechanics need to know that all the systems (pressures and temperatures, etc.), are operating correctly so that the car can be prepared for the next run. The engineer wants to

examine the car's dynamic behaviour, and the driver wants to look at the times and the rate of time slip to see where improvements can be made. The bullet points bring some order to the conflict.

IS EVERYTHING FUNCTIONING PROPERLY?

Whenever the car has been out, the first screen to look at should be the engine vital signs. In this way, if there's any remedial work needed it can be planned and started immediately. We're interested in engine rpm, pressures and temperatures, and battery voltage. Most software will allow you to bring together all the logged data onto one screen and observe the patterns. It's best to use time on the X axis rather than the more normal distance. In this way, the history of the session is obvious. Chart 6.01 shows both options, but the upper trace is shown against time, and emphasizes the warming up period and the amount of time spent in the pits.

Check the rpm trace to see that

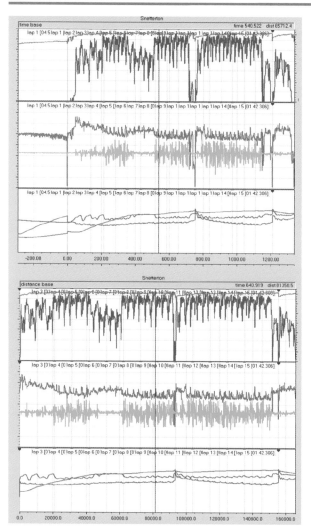

6.01. A quick engine health check shows that all readings stayed within range through the session. The upper chart is plotted against time, and gives a clearer picture of the session than the lower one, which is against distance.

the engine has not been over-revved, and look at the temperatures and pressures to see that they are within normal operating limits. Generally, the temperatures will increase throughout a run, and the engine oil pressure will fade slightly. Fuel pressure should be steady throughout and you should be checking here to see that there are no fluctuations, especially towards the end of the run.

Strip charts are not the only way to asses engine health and a useful quick check can be obtained from a pre-configured channels report. Chart 6.02 shows a typical channel report. The idea is that you list rpm, pressures, and temperatures and show the minima and maxima so you know that everything stays within bounds. This provides the absolute check giving precise numbers rather than the overview that you get from the strip chart. They can be used together to get a full picture.

Finally, at this preview stage, if you log brake pressures, it's worth taking a quick look at an XY plot of brake pressures. This shows the front pressure on one axis and the rear pressure on the other. A system that is working properly will show a tight grouping of the data into a straight line. Any problems will show up as a scattered plot and this might simply be mechanical inefficiency in the system (balance bar or valve not working well, or sticking pistons) but it might also be the first indications of a leak. If you're lucky, this can give an early warning of an impending problem before it turns itself into a major issue. The Oulton trace on Chart 6.03 was an after the event review/investigation which found a hose with a tiny split in it. This chart is now part of our post-run routine.

WHAT NEEDS LOOKING AT?

Once the care and maintenance is under way, we can turn our attention to where we can improve performance. This should always involve the driver because we'll only gain a full understanding of how to do so by using the data and the driver's experiences together. The best starting point for this is to overlay a number of laps to look for inconsistencies. Bearing in mind that time is severely limited, whether it's a test or a race day, we need to be able to home in on problem areas quickly. For the handling, this is best done by plotting the best

Lap	min	max	time of max	avg	max	avg	max	avg	min	max	min
	Engine	Engine	Engine	Oil Temp	Oil Temp	Water Temp	Water Temp	Throttle	Oil Pres	Oil Pres	Datalogger_Temp
run 1 - lap 4 06.08.966	0	6310	368.200 [s]	27	30	49	56	17.9	-0.9	6.9	-273
run 1 - lap 5 01.16.826	3625	6709	9.250 [s]	32	35	52	55	63.9	3.8	6.1	19
run 1 - lap 6 01.16.472	4099	6792	9.000 [s]	38	42	56	58	63.2	3.8	5.9	20
run 1 - lap 7 01.16.545	3462	6732	8.950 [s]	44	47	59	60	62.5	3.5	6.8	20
run 1 - lap 8 01.15.929	3888	6826	9.250 [s]	48	51	61	62	64.0	3.1	6.8	20
run 1 - lap 9 01.16.441	3876	6815	67.250 [s]	52	54	62	63	64.2	2.5	7.0	20
run 1 - lap 10 01.15.595	4006	6763	8.150 [s]	55	56	63	64	65.8	2.8	7.0	21
run 1 - lap 11 01.15.210	4311	6680	7.900 [s]	56	57	63	64	68.3	3.0	6.7	21
run 1 - lap 12 01.15.595	4115	6941	36.800 [s]	57	58	64	64	65.4	2.7	6.4	21
run 1 - lap 13 01.15.262	4008	6685	8.650 [s]	58	60	64	65	65.1	3.0	6.3	21
run 1 - lap 14 01.15.022	4145	6786	8.650 [s]	59	60	64	65	67.5	2.8	6.8	21
run 1 - lap 15 01.15.241	4111	6876	1.550 [s]	60	61	64	65	66.0	2.7	6.6	21
run 1 - lap 16 01.15.783	4239	6668	8.600 [s]	60	61	64	65	65.4	3.0	6.3	21
run 1 - lap 17 01.15.283	3938	6712	7.700 [s]	60	61	64	65	67.0	2.8	6.4	21
run 1 - lap 18 01.16.138	3954	6763	8.650 [s]	60	61	64	65	65.9	2.9	6.0	21
run 1 - lap 19 01.16.524	4068	6765	9.050 [s]	60	61	64	65	61.7	3.1	5.8	21

6.02. Most software has the ability to provide minimum, maximum and average data.

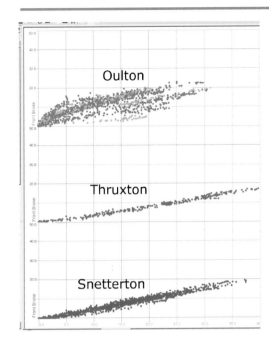

6.03. The bottom two traces indicate a reasonably well sorted brake mechanism; the top one shows that work needs to be done.

five or six laps on to one chart. You can use any variable, but speed, rpm and lateral g are the most rewarding. This gives you a view of how consistent the driver is. For some parts of the track, the data will be nice and orderly and all the laps shown will nest neatly on top of one another. The driver has hit a routine that gets through the corner without any drama. Whether this means that there is unused performance in the car will be looked at after the problem areas have been dealt with.

The problem areas show up as parts of the track where there is a spread of data and the traces are all over the place, which, in turn, indicates that things were not so easy. This is the first area to direct our attention to. For whatever reason, the driver is unable to be consistent here, and it's important to find out why. It might be a matter of technique, or commitment, a chassis problem, or a combination of factors.

Chart 6.04 shows five laps of Pembrey, and it's clear that Turn 1 is pretty well under control, although judging from the fact that the speed stays consistent and the rpm trace varies greatly, it's apparent that the driver has tried taking the corner in different gears. This would flag up one area for investigation – use the mph/1000 channel to identify which gears were used, and use the time slip

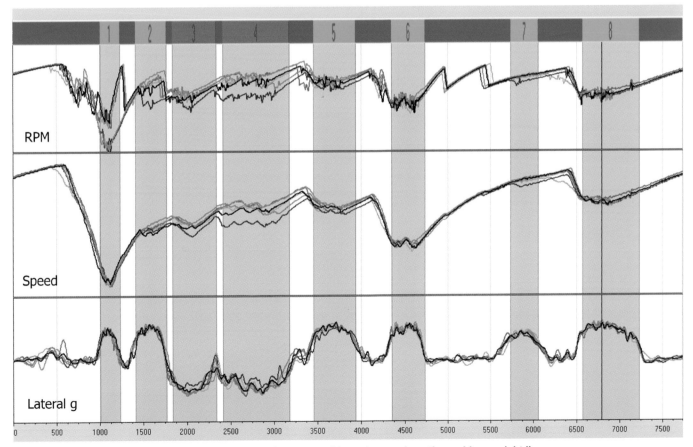

6.04. Overlaying traces from several laps shows where the problems might lie.

trace to identify whether here was time to be made, especially if a different ratio was fitted.

The rpm and speed traces from Turn 2 are messy, but again this seems to be down to gear selection again, because the Lateral g is fairly consistent, so there doesn't seem to be a handling issue there. Turns 3 and 4 definitely look like they would reward some effort. For some reason, all three traces are inconsistent. It could be that the driver has not yet mastered the two turns, or it could be that the handling leaves something to be desired. It could also be that after two right-hand turns lasting about 12 seconds, the left tyres are hot and the rights have cooled down so don't give the grip they should. Turns 6, 7 and 8 are all pretty well under control so, after dealing with Turns 3 and 4, Turn 5 would be the one to look at.

Now we have a set of priorities for investigation. The driver will be able to shed more light on the matter, and might have explanations for the

inconsistencies. Traffic is unlikely to be an explanation if the number of laps in the plot was large, but it's a possibility. If our driver has come up with a new approach to a part of the track that makes him or her significantly faster there, then there's a good chance that other competitors will get in the way. Lack of self belief or confidence in the car should be considered carefully. The driver's reasons must be analysed and, if the car's at fault, a solution must be found. Bearing in mind the old adage about not changing a car between qualifying and the race, it might not be appropriate to try it today, but it needs noting.

Often a driver will perform better at some corners than others. If there's similarity between the types of corner, you should explore reasons why this should be. It could be psychological, such as the driver having difficulty in committing to an apex which cannot be seen, or the proximity of the barriers. It could be that the car is upset by bumps

or a downhill braking area, but whatever the reason is, the questions need to be asked and explanations or fixes found.

SIGNS OF THE TIMES

The next step is the split times which give us the same information about the areas of inconsistency but in more detail. The first thing to look at is whether the best times were clustered in one area of the table. For amateur racers, the tendency is for the best times to come towards the end of the session as the driver builds up speed. If they are clustered earlier in the session then the likely explanations are driver fatigue or the car losing its edge – generally tyres, but it could be dampers or the engine losing power.

Much emphasis is placed on the fastest theoretical lap – but, in reality, we should be concentrating on the fastest rolling lap. This takes the fastest sequence of sectors irrespective of where they start. So it might be that Lap 4 was pretty messy in the first part, but

			1		2		3		4		5		6		7		8		
		histo	histo	histo	histo	histo	histo	histo	histo	histo	histo	histo	histo	histo	histo	histo	histo	histo	
Absolute split times																			
Pembrey	run 4 lap 2	8.128	4.655	2.308	4.146	0.738	5.546	0.774	7.460	2.211	4.218	3.546	4.457	8.416	2.451	3.800	5.556	3.650	01.12.060
Pembrey	run 4 lap 3	7.214	4.091	2.242	3.525	0.620	4.755	0.638	6.378	2.040	4.012	3.403	4.195	8.058	2.156	3.517	5.097	3.465	01.05.406
Pembrey	run 4 lap 4	7.412	3.923	2.150	3.427	0.612	4.381	0.601	5.953	2.019	3.951	3.377	4.261	8.079	2.154	3.414	5.015	3.456	01.04.184
Pembrey	run 4 lap 5	6.938	4.185	2.285	3.476	0.622	4.567	0.613	6.409	2.081	4.337	3.491	4.668	8.258	2.167	3.435	5.179	3.499	01.06.209
Pembrey	run 4 lap 6	7.319	4.108	2.121	3.376	0.610	4.431	0.599	6.058	2.009	3.888	3.438	4.433	7.915	2.130	3.388	5.079	3.408	01.04.310
Pembrey	run 4 lap 7	7.200	3.858	2.196	3.469	0.613	4.471	0.614	6.422	2.133	3.900	3.415	4.360	8.892	2.194	3.427	4.990	3.470	01.05.624
Pembrey	run 4 lap 8	7.236	4.088	2.149	3.370	0.596	4.337	0.587	5.905	1.960	3.865	3.418	4.320	7.859	2.127	3.336	4.913	3.429	01.03.496
Pembrey	run 4 lap 9	7.114	3.869	2.170	3.459	0.616	4.349	0.595	5.874	1.960	3.833	3.442	4.430	7.891	2.321	3.802	6.318	3.717	01.05.760
Pembrey	run 4 lap 10	8.219	4.869	2.251	3.799	0.667	4.855	0.669	6.995	2.091	4.168	3.412	4.326	7.915	2.125	3.319	5.051	3.470	01.08.201
Pembrey	run 4 lap 11	7.156	4.099	2.125	3.525	0.628	4.468	0.662	6.633	2.044	3.985	3.456	4.307	7.842	2.118	3.286	4.877	3.370	01.04.581
Pembrey	run 4 lap 12	7.140	3.842	2.194	3.442	0.621	4.739	0.664	6.840	2.105	3.935	3.424	4.290	7.807	2.184	3.379	4.967	3.416	01.04.988
Pembrey	run 4 lap 13	7.185	4.011	2.093	3.460	0.618	4.478	0.603	6.032	2.003	3.845	3.470	4.392	7.741	2.125	3.431	4.979	3.426	01.03.893
Pembrey	run 4 lap 14	6.997	4.064	2.086	3.332	0.596	**4.283**	**0.581**	6.085	2.017	3.700	**3.251**	4.196	7.883	2.114	3.284	4.913	3.406	01.02.787
Pembrey	run 4 lap 15	7.298	3.915	**2.050**	3.487	0.629	4.524	0.620	6.102	1.982	3.801	3.382	4.345	7.781	2.103	3.262	4.919	3.389	01.03.590
Pembrey	in 4 lap 16	7.082	**3.831**	2.141	**3.277**	**0.584**	4.397	0.596	**5.794**	**1.905**	3.704	3.332	4.213	**7.655**	**2.085**	**3.233**	4.896	3.446	01.02.172
Pembrey	run 4 lap 17	**6.897**	4.485	2.083	3.477	0.624	4.402	0.592	5.957	1.957	**3.694**	3.286	**4.193**	7.682	2.113	3.278	**4.784**	3.304	01.02.808
Pembrey	run 4 lap 18	6.900	4.031	2.110	3.368	0.599	4.394	0.595	5.809	1.929	3.777	3.311	4.240	7.706	2.097	3.291	4.798	3.364	01.02.317
minimum value		6.897	3.831	2.050	3.277	0.584	4.283	0.581	5.794	1.905	3.694	3.251	4.193	7.655	2.085	3.233	4.784	3.304	
maximum value		8.219	4.869	2.308	4.146	0.738	5.546	0.774	7.460	2.211	4.337	3.546	4.668	8.892	2.451	3.802	6.318	3.717	
average value		7.261	4.113	2.162	3.495	0.623	4.552	0.624	6.277	2.026	3.918	3.403	4.331	7.964	2.163	3.405	5.078	3.452	
std deviation		0.373	0.294	0.075	0.201	0.035	0.301	0.048	0.471	0.079	0.183	0.075	0.123	0.313	0.092	0.168	0.365	0.100	
Theoretical best lap																			
Pembrey	best	6.897	3.831	2.050	3.277	0.584	4.283	0.581	5.794	1.905	3.694	3.251	4.193	7.655	2.085	3.233	4.784	3.304	01.01.399
Best rolling lap																			
Pembrey	run 4 lap 16		3.831	2.141	3.277	0.584	4.397	0.596	5.794	1.905	3.704	3.332	4.213	7.655	2.085	3.233	4.896	3.446	
Pembrey	run 4 lap 17	6.897																	01.01.986

6 05. Sector times report.

got better towards the end. In Chart 6.05 lap 16 was the fastest actual lap except that Sector 1 was slightly slow. If Sector 1 from lap 17 was included, the fastest rolling lap is a couple of tenths faster. That represents what the car and driver are actually capable of, even though it does not coincide with the official timing, and will not earn you the lap record.

ABSOLUTE VALUES

Looking at overlays and split reports gives us some direction for further investigation based only on inconsistencies. We should also be concerned with absolute values. If the car generally achieves values of 1.6g in the corners and 1.4g under braking, we should expect to find these values in this data. A quick way to do this is to look at lateral, longitudinal and combined g all on one chart.

If the software has the option, using a full cursor that stretches the whole width of the chart is a useful technique. If you haven't seen this feature, it puts cross hairs on to the point where the cursor meets the trace. In this way, you can look down to see what the other traces are doing or across to see how other values compare with the one where you have the cursor. This is used on the lateral g trace in Chart 6.06. The cursor measures the peak value, a scan across the chart shows that the peak value is being achieved everywhere except Turns 1 and 7. You'll have to make allowances if aerodynamic effects are significant. Turn 1 is already on our list for further thought, Turn 7 was not known about.

Investigation (see Chart 6.07) using the speed trace shows that the speed keeps increasing all the way through

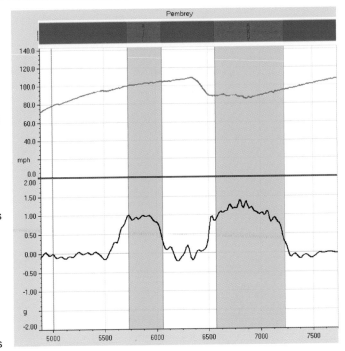

6.07. Turn 7 fails to generate high cornering forces simply because the car is not fast enough to do so.

Turn 7. It's not the car or the driver that is responsible for the lack of lateral g, it is the lack of power. If the car cannot generate any more speed, the driver cannot generate any more g force.

The single most important fact with this particular set of data is that, no matter what else we read into it, lateral g only reaches about 1.5g, and so falls short of the expected 1.6g. Looking at the braking levels would give us a clue as to the overall level of grip – if long g was down by a similar percentage, then it might be safe to assume that the track is less grippy than the others, or that this set of tyres is not performing up to expectations. Before buying a new set it would be wise to see how the other teams are doing.

Again, we don't only have the charts to look at; we could use a channel report to summarise peak and average values to give us a feel for what's happening. Channel reports can be particularly useful for comparing performance from

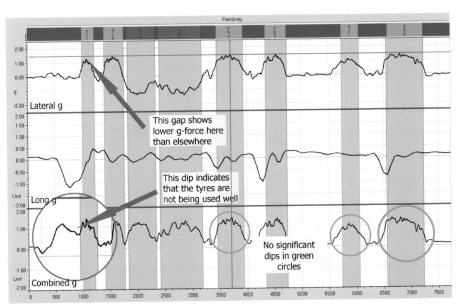

6.06. Showing all the g forces on one chart gives a feel for whether acceptable absolute values are being reached.

one session to the next. Showing the average speed of laps brings home to you just how fine the margins are that we're looking for – the difference between good and best is small. A series of hot laps should exhibit average speed differences of no more than one or two miles per hour, and if they do, then either the driver or the engineer needs to do a better job.

THE DRIVER'S PERSPECTIVE

Now it's the driver's turn to consider the data simply from the driving point of view. Using the knowledge gained from the analysis so far, it's easy to see that there's room for improvement in the early part of the lap, but the big issue is whether or not the overall lack of performance is due to the car or the driver. It's important not to get sucked into over-driving just because you're not hitting the times and expected values.

The starting point must be to explore the issues raised in the first examination of the data. This is pretty tedious because any competitive driver will want to be looking where to find the extra tenth or two. The lap and sector times will have given a huge hint where to look, but it will have to wait for later because doing things the engineer's way will make sure that there's time to make whatever changes are necessary.

In the data used here, there's an issue surrounding the choice of gear for Turn 1 and possibly Turn 2, so that can be looked at separately. The next stage is the business of improving the first half of the lap.

Chart 6.08 shows the data necessary for understanding the gear choice; the green lines show the use of 1st, and the blue and purple the use of 2nd gear in Turn 1. In terms of speed, the blue line is best (it represents the baseline of the time slip trace) so staying in second gear gives a slight time advantage. This advantage is pressed home by not having to change gear in the middle of Turn 2. The rpm figure drops well off the power curve in Turn 1,

so it looks like there would be benefits in fitting a lower second gear as far as this corner is concerned. This might hurt at other points on the track, so this opens up another investigation, this time into gearing (discussed in Chapter 8).

Chart 6.09 shows speed, cornering force, and driver inputs for the three fastest laps. But here we are focussing on the first half of the lap and trying to get a measure of performance. The stark fact to be faced from this set of data is that Lateral g is 0.3 to 0.4 behind the target for this car in Turns 3 and 4. The early hypothesis of it being to do with cold tyres does not hold water because the second turn generates less lateral g than the first. The driver has to accept some of the responsibility for this, and think hard about the way these corners are being tackled. The engineer is not completely out of the firing line, because elsewhere on the track the values are at least 0.1g down. You should never restrict yourself to just the fastest laps because there are often little cameos of

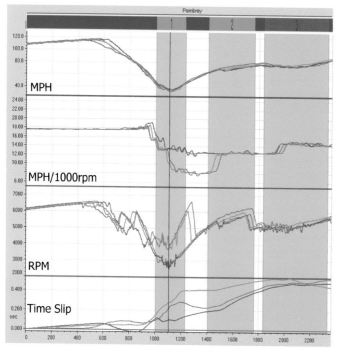

6.08. The first corner gear choice issues.

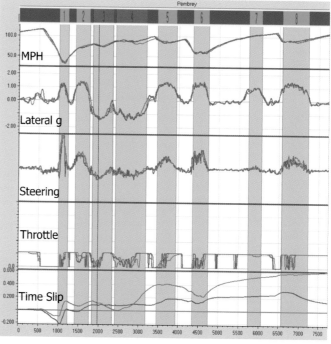

6.09. Looking at handling issues for the rest of the lap.

speed or bravery tucked away in much slower laps as well.

Time now for the driver's chance to look for the tenths and hundredths that will improve the next run. The trick here is to use the time slip line to look at as many laps as possible to see where the time is to come from. Whenever the line is heading down the chart, time is being saved. The red ellipses on Chart 6.10 show the three instances where this happens. The two on the left need to be used with caution. They are in the braking zone and the fall in the line is usually followed by an equal or greater climb. This is the easy error of in fast, out slow. So although there might be something to be gained from late braking, its effects on the later phases of the corner must be considered. On the other hand, the right-hand ellipse shows time being gained in the final phase of the corner on the red and purple laps, this is translated into time saved on the following straight as well. This is the real area to work on.

Chart 6.11 is based on issues identified in Chart 6.10. It is zoomed in on Turns 7 and 8 and only the red, blue and the purple laps have been retained. Here, there are small but real time savings to be had (in fact, about 0.15sec). The steering trace is interesting. The red and the purple traces show that the driver is using less steering angle on these laps for very similar lateral g, and the conclusion that you draw from this is that the car is oversteering slightly. This is confirmed by the slight jaggedness of the trace indicating rapid adjustments to the wheel position. Although it is not very clear from this screenshot, the driver is also being more aggressive with the throttle. It certainly seems that this is a corner where the car can be bullied a bit more.

Although this has only been a case study of one session, it is the principles that matter. Time is limited, so the health of the car is the first thing to check. This maximises the time available for any remedial action. The next step is to get a feel for the overall level of performance by looking for inconsistencies and at absolute values. This examination of the data might prompt decisions about changes to the car, so this should be scheduled next, again, to make time available for changes. Finally, the driver needs to evaluate his or her performance by a structured look through the data and with the minimum amount of self deception.

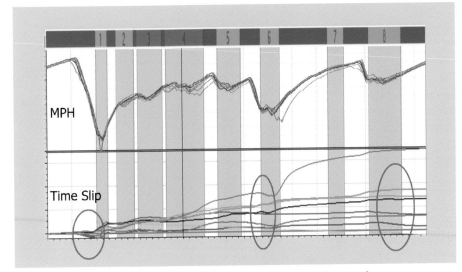

6.10. The red ellipses show where time *might* be saved.

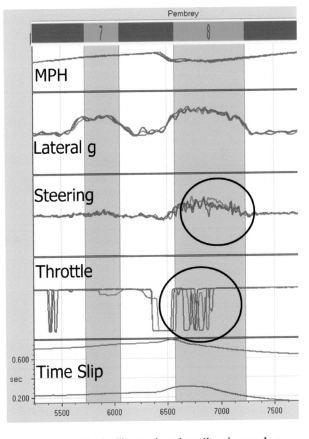

6.11. The black ellipses show how time *is* saved.

Chapter 7
Handling & steering

Let's start by admitting that there can be a gap between the expectation and the reality of what a logger can do. New users expect to be able to plot a few key variables (steering, throttle and lateral g) and see immediately what's going on. It isn't impossible, of course, but it takes a bit of effort and experience. With the limited equipment at our disposal, we're never going to get definitive answers, but we can certainly learn enough to gain a competitive advantage. With a little extra thought we will be able to go much further into secrets hidden in the data. The fact is that the logger will only show what is happening to the key variables. Success depends on how you pull this information together and interpret it.

This chapter shows how you do this. The first stage is to look at some overall indicators of how well the car handles in general, then look at the relationship between the car, the driver, and the track, and then move on to a more complex maths-orientated approach. Once we have an

understanding of how the engineering analysis can work, we look at shortcut methods that give excellent results for not too much effort. Finally, we take a quick look at what the future holds with GPS-based analysis.

A top level professional team with access to more channels of information should get much closer, but, in doing so, would probably log:

• damper movement
• lateral g at the front and rear axles
• ride height
• tyre temperatures

The best teams will be able to bring all of this data together to improve their understanding of what the car's doing, but there's also reason to believe that many professional teams have more data than they can cope with. Compared to this, the typical amateur team is stumbling along in the dark, but even our limited range of sensors can provide plenty of insight. The key is to use the

data analysis and driver's interpretation together to help clarify the situation.

The logger is an important source of information, but so is the driver, and it would be foolish not to make the fullest use of both. This chapter begins with the easy, intuitive ways of interpreting the data, and goes on to look at more rigorous approaches that involve calculating the corner radius from speed and lateral g data, and using this to work out the amount of steering that should be needed at any point on the track. By comparing actual steering with required steering, we can get a pretty good idea of the balance of the car. Unlike the highly-resourced professional teams, what we won't be able to do is say with certainty that the right rear damper needs its low speed rebound setting stiffening, for example.

OVERALL MEASURES
Although it's not particularly exciting, it is possible to get a broad understanding of how well things are going by looking

Lap	max	avg	max	avg	min	avg	max
	Engine	Speed_1	Speed_1	Throttle	Acc_1	Acc_1	Acc_1
run 1 - lap 3 03.12.343	6202	37.1	81.0	19.2	-0.89	0.10	1.19
run 1 - lap 4 01.31.826	6540	77.6	120.6	52.4	-1.21	0.22	1.29
run 1 - lap 5 02.01.441	5792	58.2	102.5	21.3	-0.95	0.14	1.02
run 1 - lap 6 02.03.517	6621	57.4	100.2	24.4	-1.20	0.15	0.98
run 1 - lap 7 01.20.039	6569	87.6	120.4	58.1	-1.13	0.30	1.53
run 1 - lap 8 01.19.278	6725	88.4	123.7	58.8	-1.30	0.30	1.54
run 1 - lap 9 01.19.486	6735	88.0	124.1	61.7	-1.22	0.30	1.51
run 1 - lap 10 01.20.154	6732	87.3	123.4	63.1	-1.28	0.29	1.52
run 1 - lap 11 01.19.351	6529	88.2	117.4	60.2	-1.14	0.30	1.45
run 1 - lap 12 01.18.756	6692	88.9	123.0	66.6	-1.36	0.32	1.53
run 1 - lap 13 01.18.276	6888	89.5	123.2	64.5	-1.30	0.32	1.53
run 1 - lap 14 01.17.400	6710	90.5	123.4	66.9	-1.33	0.32	1.69
run 1 - lap 15 01.18.537	6738	89.1	124.1	64.7	-1.38	0.30	1.61
run 1 - lap 16 01.38.169	5801	71.3	106.8	33.6	-0.98	0.20	1.29
run 1 - lap 17 02.03.360	5917	57.1	89.5	19.5	-0.88	0.12	1.02
run 1 - lap 18 02.22.116	5335	50.1	81.0	17.0	-0.90	0.10	0.96
run 1 - lap 19 02.22.898	5401	49.6	77.6	19.1	-0.96	0.10	0.96

7.01. The software should show a channel report of the key data.

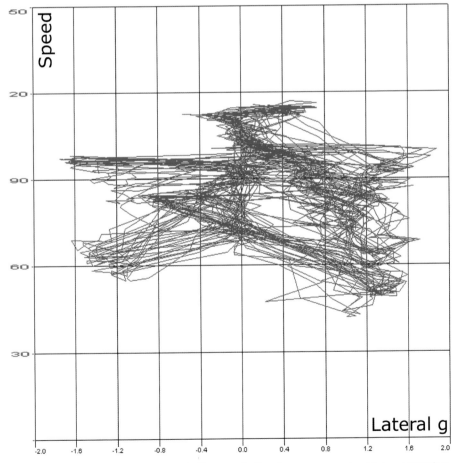

7.02. Lateral g vs speed; this chart indicates whether the car generates higher cornering forces at higher speeds.

at the data in total, and getting a feel for the overall statistics. For example, using a channel report that shows peak lateral g will tell you whether you're in the expected operating range for the tyres and vehicle type. Average lateral g can also be informative as a measure of comparative performance at the same track on previous occasions. Each track has its own unique value for average lateral g, but it allows you to compare this outing with previous ones.

XY charts can also shed light on handling. The friction circle (long g vs lateral g) and steering vs lateral g are covered in Chapter 11 because they give the driver some insights into his or her performance, but plotting lateral g against speed can also give us an idea of how we're doing. On Chart 7.02 lateral g is plotted on the X axis and speed on Y. Left-hand bends are shown towards the left and higher speeds on the upper part of the chart. The traces generally stretch out to + and - 1.6g, which is satisfactory for this type of car. We can also be fairly comfortable with the aerodynamics because, as the speed increases, so does the cornering force. The effect is very mild, but the car has no wings, just a reasonable body shape.

The two left-hand bends are quite distinct, one being taken at just over 60mph and the other at around 100. The right-hand bends are less distinct, but the overall impressions to be gained are that the slower the bend, the lower the lateral g. This indicates a mild level of aerodynamic download. Generally this type of chart should roughly resemble an inverted triangle where there is downforce, and be more square-shaped where there is none. The shape to watch out for is if the triangle has a flat base and tapers towards the top. This would indicate that high speed bends were causing more difficulty than the slower ones. On shorter tracks, where the individual corners can be identified, this can be a good indicator of driver consistency. The chart shows where one corner is generating different lateral g to the others. This can be misleading if there are 'non-event' corners, but generally it's a useful measure.

STRIP CHARTS FOR HANDLING & STEERING

Although it's possible to do all sorts of mathematics on the steering data (and this follows later in the chapter) it's well worth looking at the raw data because there's a lot of interesting information hidden just beneath the surface. At the basic level, it's possible to simply read the various traces and to get some understanding of vehicle behaviour from that. An oversteering car shows a jagged lateral g and steering traces with the driver applying lock and then reducing it as the back of the car starts to step out of line. Once this is corrected, more steering is applied and the sequence starts all over again. The lateral g trace tells the same story – there will be increases in lateral g as the steering forces build up, followed by a reduction as the steering is corrected. Both traces will have humps and hollows, but those

in the lateral g trace will be more muted than those in the steering.

Understeer is not so dramatic, but is still easily recognizable. There might be the desperate rally style 'chuck' into the corner with a flick or double flick on the steering, to unsettle the back of the car to try to provoke it into an attempt to overtake the front. This desperation would then be followed by some sort of corrective action to secure a reasonable progress through the rest of the turn. A good engineer or driver ought never to see this sort of extreme measure. Much more likely would be plenty of steering to get the car turned in, followed by a reduction in steering lock as the car and the driver come to terms with the corner, followed by some more lock to finalise the corner and get on with the lap. Lateral g would climb slowly and fall away in a similar fashion.

Lateral g

The starting point is the lateral g trace, and this section looks at this data in isolation. This is a bad idea and is only done because some systems don't provide steering or throttle data for further back-up. The shape of the lateral g trace will give us some clue as to the line taken. It will also help us see how the car is handling. Be warned, though, the effects are very subtle, and it's easy to confuse understeer with oversteer. It's when you get to this stage in the interpretation process that you'll either regret the lack of a steering trace or be pleased that you made the effort.

The easier of the two handling symptoms to recognise is oversteer. A sudden and significant loss of rear-end grip during the corner shows up as a rapid dip in the lateral g trace, and driver's response causes a sudden climb back to former levels. The dip and climb are so close together that they look like a spike in the data. It's so visible that it cannot be missed. This can be seen in

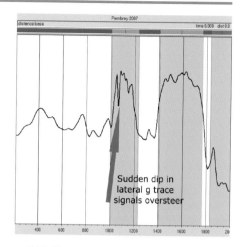

7.03. The sudden and short-term dip in lateral g indicates an oversteer incident.

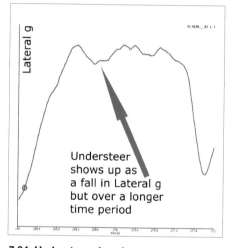

7.04. Understeer also shows as a reduction in lateral g, but lasts significantly longer.

Chart 7.03. The 'right-on-the-edge' sort of oversteer can also be identified simply by reference to the lateral g trace on its own. Here, the trace is jagged as the g forces fall and are dragged back by the driver's skill and daring. This shows that the driver felt the thing let go, and then took the necessary action to get it back again later in the corner.

Understeer is more subtle, and can also show itself as a sudden dip in the lateral g trace. The difference is that the dip is shallower and lasts longer. Shallower because even though

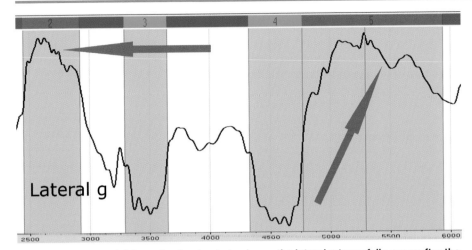

7.05. The arrows point to understeering behaviour – the lateral g trace falls away after the initial peak.

the front end of the car has let go quite significantly, there's still some residual cornering power. If the back lets go, it tends to head straight for the barrier; if the front lets go, it shows itself as an uncomfortably large increase in the corner radius. The correction lasts longer because the mid-corner correction for understeer is to lift the throttle, possibly reduce the steering and wait for the problem to cure itself. This takes more time than the deft flick of opposite lock that catches the oversteer.

The understeer shown on the first part of Chart 7.04 was sudden and significant caused by a wet track. Run-of-the-mill, lap-by-lap understeer shows up on the trace, but it's subtle and needs looking for. Reluctance to turn in shows as a slow and hesitant increase in lateral g, probably reaching peak values in a couple of steps as the driver either patiently waits or takes some more vigorous remedial action. Mid corner understeer is difficult to spot using the g trace alone; it has to be inferred from a smooth curve whose peak values do not reach expected levels. The reluctance of the car to complete the turn shows as the lateral g trace flattening out towards the end of the bend. It's the gradual

changes in lateral g that you need to look for, rather than the drama of the jagged oversteering trace. Understeer only gets dramatic when the driver is taking extreme measures to hurry the car along.

Chart 7.05 shows a typical trace from a car with understeer. In both corners the initial lateral g levels fall away after the middle phase of the

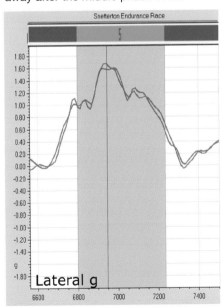

7.06. Snetterton's notorious Bomb-Hole confuses the interpretation of the lateral g trace.

corner. Again, it's the time taken for this to happen that helps distinguish it from oversteer. Typically, an event lasts more than a second. Corner entry understeer shows up in the slow build up of lateral g.

Bumps in the track surface add to the difficulty in interpretation. The clue here is that a particular handling phenomenon occurs lap after lap. It might be that the car has a handling problem, but equally likely is the fact that the track surface affects the car on every lap. The trace in Chart 7.05 would lead you to believe that there is understeer at the beginning and end of the turn, but, in fact, the dip in the track causes the driver to tighten the line significantly in the middle of the turn. Most cars show this behaviour in this corner.

Steering and throttle data

Although it's possible to interpret handling from the lateral g trace alone, it doesn't make much sense to try to do so. Combining the lateral g with steering and throttle gives us a much more complete picture, because although lateral g is the output, it's the driver that provides the inputs through throttle, steering, and choice of line. This section looks at ways of pulling these things together to understand what's going on. To repeat something that has been said already in the book, the driver must be there to add his or her insight or you're wasting a valuable resource.

Bringing together the lateral g and the steering and throttle data, and understanding their message, is pretty much a matter of common sense. Charts 7.07 and 7.08 demonstrate just how useful the steering data can be. In both cases, the lateral g on its own tells the story, but the additional steering data makes things much clearer. It is really not worth being without steering and throttle sensors, because they're

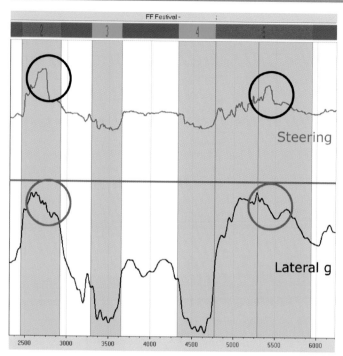

7.07. This is the understeer trace shown previously in chart 7.05 made much easier to read by the addition of a steering trace.

7.08. The sudden oversteer from chart 7.03 is also much easier to spot with a steering trace.

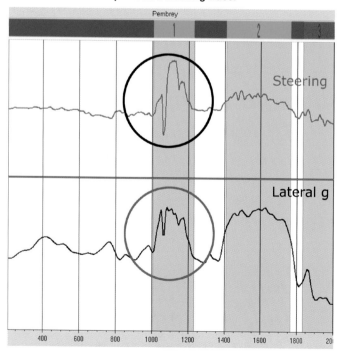

so inexpensive and easy to fit.

It is important, though, to recognise that the driver's style will have a lot of influence on how the data looks. Some drivers are naturally relaxed and wait for the car to tell them what to do. Others are more tense, feeling that they always need to be on top of the car and giving it inputs. Both styles work, but 'relaxed' is much easier than 'tense' when it comes to interpreting the data. The driver that feels the need to boss the car at all times tends to

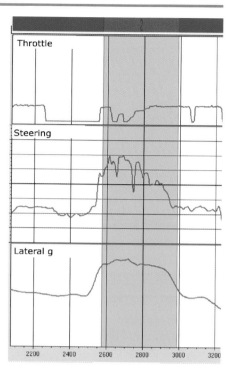

7.09. The jagged steering trace here indicates a slightly nervous oversteering car, and a driver who likes to be on the case all the time. He is also greedy with the throttle and, because of this, has to apply significant amount of opposite lock on two occasions in this corner.

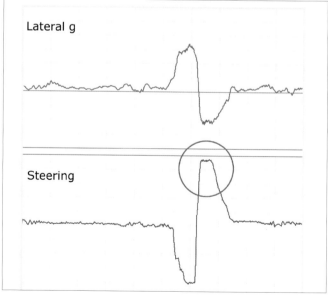

7.10. A contrasting style, letting the car do the work: note the relaxed consistency of the steering trace.

use lots of spurious inputs that show up on the data, but that really tell us more about the driver than the car.

There can also be a problem of terminology. Some people would only regard it as oversteer if the steering correction (inside the black circle of Chart 7.08) actually crosses the zero line. Some drivers regard a jagged steering trace a simply the basic business of hurrying a car round a track. They regard it as a series of minor corrections that keep everything under control and call this continuous sequence of movements of the steering wheel 'driving.' Others have different expectations, and search for an ideal where, once turned in, constant steering can be used to maintain the chosen line. To them the need to keep reducing the steering lock is evidence of oversteer. It's the improvisers versus the artists.

MORE COMPLEX APPROACHES

So, with a careful look at the data, plenty of common sense, and what the driver recalls, it's possible to learn an awful lot about how the car behaves. Simply looking at lateral g, steering and throttle helps us go a long way towards telling the story and represents the 80 per cent of the results that we get for 20 per cent of the effort. The rest of this chapter deals with more complex methods. Whether you choose to stop at this stage or take the analysis further is up to you.

Corner radius

If you want to take things further, a good starting point is the 'corner radius'. This is useful no matter how simple or complex your approach, because it makes it possible to compare different corners at different circuits. In this way we can begin to predict cornering speeds, and make gearing decisions. Knowing the corner radius can also help us understand what the driver is (or

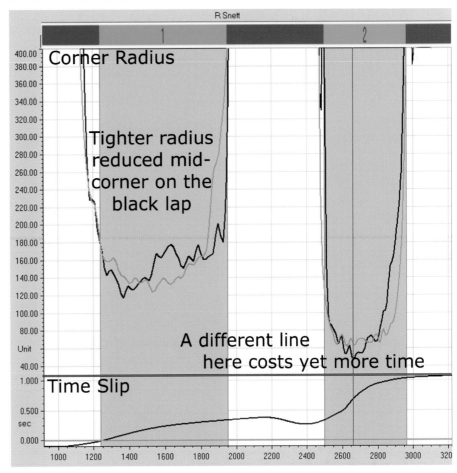

7.11. The corner radius trace shows what lines were used.

should be) doing. On a large radius bend (fast corner) all that the driver need do is lift momentarily to settle the car and then get back on the gas. In tighter turns, the throttle needs to be applied with greater care. If you're going down the full racecar engineering route, and intend to create your own oversteer/understeer channel, knowing the corner radius will be essential. Corner radius comes as standard in some software but if you have to go down the do-it-yourself route, the calculation is pretty straightforward:

Corner radius = speed^2 ÷ lateral g

The units have to be consistent (speed and lateral g both in metres and seconds or feet and seconds) and

the data that you get will be a series of enormously large numbers, representing the straight parts of the course and a few smaller numbers representing the corners. If you think about it, another way of looking at a straight is that it is a corner with an infinite radius. The very high numbers could be taken out usinga bandpass filter (see Chapter 1).

The graph will also look better if all values are converted to positive numbers by squaring them and finding the square root. In this way it's possible to plot corner radius that gives some indication of the line the driver is taking. This is dealt with in the chapter on understanding the driver. To make all the numbers positive the formula needs to be amended to:

Corner radius = square root (speed[2] ÷ lateral g) [2]

Using a bandpass filter and converting the numbers to positive values gives a trace like the ones shown in 7.11. The corner radius is a useful channel in its own right because it gives a view of the line that was taken. The logger records a very sharp turn-in as a very a steep drop in the line, and a constant radius would show up as a flat bottom to the curve. Gradual releasing of the line is shown as a less than vertical climb of the trace.

The corner radius in 7.11 is not so well filtered as the pre-calculated ones that are built into some programs, so the relationship between it and a lateral g trace can be seen. After all, the corner radius is simply speed adjusted lateral g data. There are at least two possible causes for the step up in the black trace in the middle of Turn 1. It's probably understeer-related because it mimics lateral g and has understeer's characteristic shape and the time period. So, this suggests that the effect was to reduce the corner radius as the car slid towards the outside. But this bend is, in fact, a double apex bend where the fast line is a constant radius as shown in the orange trace. The black trace might simply be that the driver allowed the car to move too far out after the first apex and had to tighten the line to hit the second apex. Again, this is an example of the need for the engineer and the driver to read the data together (and a case for including video in our analysis).

Because of the complication of converting all the values to positive and having to filter out the large numbers, some people prefer to use the inverse of the corner radius, like so:

Inverse corner radius = 1 ÷ corner radius

This has the advantage of being a

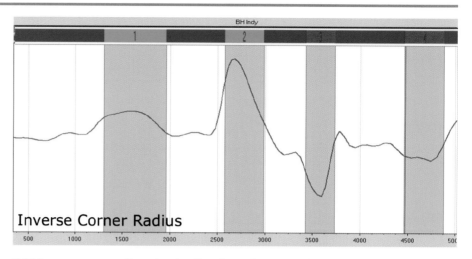

7.12. Inverse corner radius, showing Turn 1 as a fast corner in which the driver adjusts his line about halfway through, and again at the end. Turn 2 is much tighter.

lot simpler, and getting rid of the very large numbers.

Instead of a straight being a corner of infinitely large radius, and, therefore, shooting off the top of the graph, straights are calculated to be zero, and the bigger the number, the tighter the turn. The trouble is that, when measured in metres, this gives values between 0 and 0.03 so they are hardly informative numbers, although they do plot nicely. Figure 7.12 shows a plot of an inverse corner radius and, again, it gives us a feeling for the sorts of lines that are used. It's not a map of the corner, just a plot of the radius at any point in time. The classic symmetrical racing line would have a steep slope as the driver turned in, a plateau or a flat bottom as the required radius was reached and maintained, and a steep slope as the car was released on to the straight. The three bends of Figure 7.12 show something a bit different. The first turn is almost symmetrical except that the driver tightened the line very slightly, mid-corner. The second turn is shorter, with an early apex and, almost as soon as the steering had been applied it was taken off. The third turn had an adjustment to the line just after the half-way stage.

Speed adjusted steering (SAS)

One problem with trying to compare steering to lateral g is that the two do not move in unison. Without aerodynamic download, a car can be expected to reach similar levels of lateral g whatever the radius of the corner. It doesn't matter whether the corner is taken at 30mph or 130mph, without aero, the car will still pull more or less the same g forces. Steering, on the other hand, will vary with speed. To get round the 30mph bend could need all the available steering, but the 130mph bend would need very little. So, just trying to understand the handling by comparing the steering and the lateral g traces is very difficult. One approach to this problem is to adjust the steering to account for the speed, and to create a maths channel called 'speed adjusted steering'. There are, apparently, many ways in which this can be done, but the one that most people use is:

Speed adjusted steering = steering x (speed[2] ÷ √speed)

Or in the arcane language of a maths channel, this will look something like:

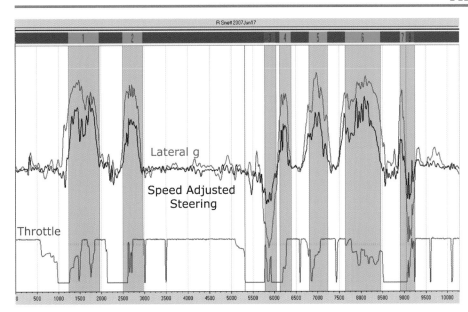

R.Snett 2007Jun17

Lateral g

Speed Adjusted Steering

Throttle

7.13. Speed adjusted steering. Choose one corner where the car is well balanced, and make a judgement on the others.

Speed adjusted steering = steering*(speed_1^2)/(sqrt(speed_1))

This has the effect of changing the steering trace by 'bulking-up' the amount of steering displayed in the higher speed corners so that it's more in line with the lateral g or the steering trace. The formula certainly adjusts the steering for speed, but whether it does it correctly is another matter.

This doesn't give definitive answers, because you need to use judgment in interpreting the trace. First, you identify a corner where the car was well balanced. From this we can identify the relationship between speed adjusted steering and lateral g. It might be that with the current scaling, the SAS was equal to lateral g and tracked it pretty well exactly. On the other hand, it may be that the SAS trace was only half the height of the lateral g trace. Whatever the relationship is, you would expect to see it repeated in all the turns where the car was well balanced. If we take the case where the SAS trace was about half the height of the lateral g trace, then, when that relationship

changes, it tells us something about the handling. If the SAS is less than half the height, then the driver is using less steering and the diagnosis is oversteer. If the SAS is taller than expected, the driver is applying more steering and we think of understeer.

In Chart 7.13, if the car felt well-balanced in Turn 2, then we have to suspect oversteer in Turn 3, and understeer in Turns 1 and 6.

How much steering is needed?

If the speed adjusted steering technique seems a little bit subjective, another approach is to compare actual steering with the required steering. Actual steering is one of the logged channels, so that's no problem. Required steering depends on the corner radius and the wheelbase of the car. This requires effort to set up the requisite channels, but the reward is a much more reliable indicator of oversteer and understeer.

From speed and lateral g we can work out the instantaneous corner radius, so we know precisely what path

the car was actually following, and if we take into account the wheelbase of the car, we can work out the amount of steering that was required for that trajectory. The difference between the required steering and the actual steering is a measure of understeer or oversteer. How much work you choose to do is a matter of personal preference, and there are quick and dirty surrogate measures that will tell you the same thing without actually putting numerical values to it.

As we've already seen, more steering is required in slow corners than in fast ones. To negotiate a bend with a 25m radius a car with a 2.5m wheelbase would need:

2.5 ÷ 25 = 0.1 radians of steering lock

To convert this to degrees we would multiply by 57.3 so the answer would be 5.7 degrees. The same car in a faster corner (say 100m radius) would need:

2.5 ÷ 100 =.025 radians (or 1.4 degrees)

For people who measure their distances in feet, roughly equivalent numbers would be a radius of 80ft and wheelbase 8ft. This would give:

8 ÷ 80 = 0.1 radians of steering angle (or 5.7 degrees) and in the faster turn, the radius would be 8 ÷ 320 = 0.25 radians (1.4 degrees)

The introduction of radians as a means of measuring angles should not cause non-mathematicians any undue concern. Instead of splitting a circle into 360 degrees this method splits it into pi times 2 or 6.28 radians. As these calculations show, this apparently clumsy method actually makes some of the arithmetic much simpler, but it leaves you with a bit of a problem. Engineers will happily use radians as a measure, but generally, degrees are much more

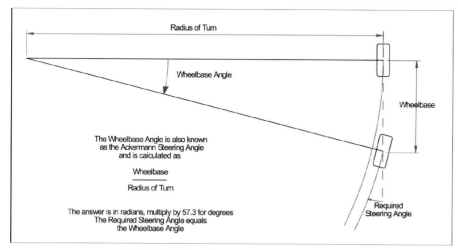

7.14. Calculating the required steering angle.

widely understood. So you have to choose between radians for easier calculations and formulae, or degrees so that you don't discourage people from getting involved with your numbers.

Required steering

As shown above, the formula to calculate the required steering angle is fairly straightforward so long as we bear in mind the need for consistent units (metres for the wheelbase and m/sec for both speed and lateral g, or using feet for the wheelbase and ft/sec for speed and lateral g).

Required steering angle = (wheelbase ÷ (speed^2 ÷ lateral g))*57.3

This channel is particularly useful because it deals with the actual path followed by the car, so if we compare this with the actual steering input from the driver, we have a measure of whether the car is oversteering or understeering. This only works properly if the steering channel is also calibrated in degrees of road wheel movement, because the figures will be directly comparable and, if plotted on the same chart to the same scale, we would only need to look at where the two traces did not coincide. In fact, we could go a step further and

create a handling channel calculated by subtracting one trace from the other. If your steering channel is calibrated in some other way, the section below called 'The lazy way to a handling channel' gives an effective workaround.

Some software has this channel or a version of it as a standard feature, but this relies on the user configuring the logger and the software to use channel names and units that the software requires. It seems much safer to turn your back on the pre-fabricated version and work out your own channel from first principles. In that way you see what it is that you're measuring when you claim to be measuring understeer.

Using someone else's standard handling channel that subtracts the required steering from the actual steering and shows the difference as understeer or oversteer is undoubtedly more sophisticated, but is one step further removed from the raw data and is not as understandable. Outsiders tend to regard this sort of channel as some form of special magic and then either dismiss it as such or trust it blindly. In either case, they have suspended their critical and analytical faculties which is exactly what they should never do. The step-by-step, do-it-yourself approach keeps you more in touch with reality.

Do not trust this calculation completely. There is a small snag hidden in the logic because of the tyre slip angles. We record what the steering on the vehicle does but the tyres operate at a slip angle, which means that there's a difference between the direction that the road wheel is heading in and the direction that the tyre contact patch is heading in. Theoretically, oversteer is when the rear slip angles exceed the front slip angles. By comparing required steering with actual steering we're glossing over the fact that the slip angles will differ when the car is understeering or oversteering. The measure still seems to work pretty well, and reflect the driver's interpretation so that it gives us a reliable way of understanding handling.

The lazy way to a handling channel

The previous section is all well and good if the steering is calibrated properly but there is a shortcut. All that we need is for our steering data to be in the form of zero for straight ahead, have negative values when the lateral g goes negative and positive values when the lateral g goes positive. The trick is to plot the two channels on the same chart. At first the two will exhibit similar shapes but not match up at all but by scaling the required steering channel, a really good fit should be possible.

We can use slower laps to provide a baseline for our data. All that is needed is that the driver does a slow lap to give us an idea of what the charts should look like in a neutral handling car. No driver will like this, because they all want to be 'on it' from the moment the car leaves the pits. Inevitably, since most amateur drivers also pay the bills, and track time is so limited, laps driven well within the capability of the car tend to be few and far between. Safety car periods provide one realistic opportunity; just one lap of any caution period can provide valuable baseline data before the tyres have to be brought back to temperature. The slowing down lap at the end of a session provides another opportunity. What is needed is something well within the capabilities of the car so a cornering speed that is about 80 percent of what would be done in the heat of competition is about right. We're not talking about snail's pace, just taking proper racing lines and not sliding the car around.

To fit the baseline data, we need to overlay the actual steering and the adjusted steering channels. Then the scale of the required steering needs to be adjusted so that the two traces coincide as closely as possible. This will take a few attempts, but by changing the scale on the required steering trace, you should end up with something that coincides pretty closely with actual steering. Chart 7.15 shows the results of this for a pace car lap. Once you've hit on an effective scaling that matches steering and required radius, it will generally work for other circuits and outings. Once satisfied with this, you can turn to the competitive laps and interpret the chart.

Chart 7.16 shows the traces for a race lap, and it shows handling characteristics quite clearly. Where the measured steering trace is outside the required steering trace, the driver is applying more steering than the maths say that is necessary. This might be a quirk of driving habit, but it is much more likely to be an indicator of understeer. When the measured steering channel is inside the required steering trace, this is evidence of oversteer. Once the car is turned in, the back of the car (with or without assistance from the gas pedal) is moving further than the corner radius requires.

All of this is only evidence not proof, and will need to be considered in conjunction with the lateral g and throttle position traces and, as always, discussions with the driver. The comforting thing is that the two traces come from two completely different sources and match up very closely. Required steering is calculated using speed and lateral g, and steering is measured directly from the car.

Handling from GPS data

One of the techniques used by engineers who are developing self-steering vehicles is to compare GPS derived data about lateral acceleration and yaw, with the data from onboard sensors, and use the differences to help understand what the vehicle is doing. At its simplest, a comparison between the lateral acceleration measured by an onboard sensor and by the GPS data can tell us how the car is handling.

The principle is that the GPS data gives an analysis of the car's speed and trajectory, and from this we can calculate lateral g. In this way the GPS provides an overview of the lateral acceleration that the car is experiencing. Back on board the car, its own lateral accelerometer located in the middle of the car measures the lateral g directly. If the car is handling neutrally (is not oversteering or understeering) the two sets of data will coincide. But if the car is oversteering, then there's a small inaccuracy in the onboard measurement. Instead of being at right angles to the direction of travel, the accelerometer is skewed round at the angle of drift that the car is experiencing. So, if we look at the difference between the two lateral g figures, we have an instant indication of the extent to which the car is oversteering or understeering.

To create an understeer/oversteer channel, we simply need to subtract one lateral g figure from the other. Sadly,

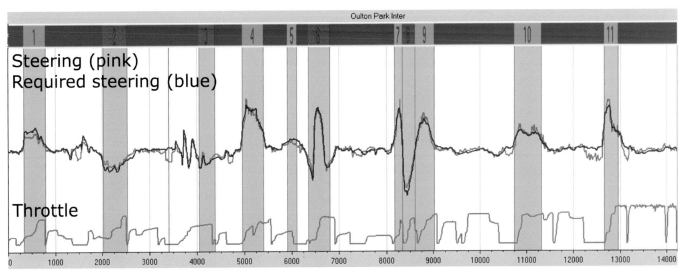

7.15. A lap behind the Safety Car gives a good basis for adjusting the required steering channel.

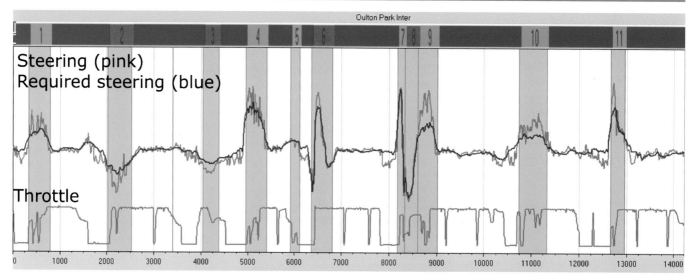

7.16. Four laps later, the traces are significantly different to those in 7.15.

although the current GPS-based loggers are capable of doing these calculations, they do not yet provide enough accuracy to make this technique useful. There's not sufficient similarity between the two sets of data to give a reasonable indication of handling, although, as the accuracy of commercial data logging systems increases, so will the chances of this promising technique. The GPS systems used in the self steering vehicles use two antennae and the differentiated signal provides a more reliable signal on which to base the calculations.

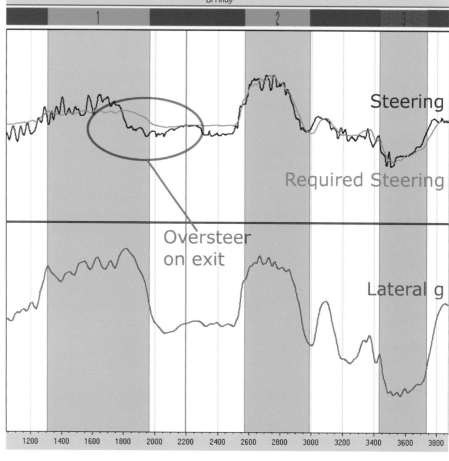

7.17. Actual vs required steering clearly shows the oversteer coming out of Turn 1.

Chapter 8
Power, gearing & traction

The purpose of this chapter is to look at how to get the best from the engine by making the best use of the available power. We can use the logger to:

• monitor engine power output
• choose the best gearing
• identify traction issues

The starting point for all gearing decisions should be the engine's torque curve, because the thrust available at the driving wheels is the torque multiplied by the gearing and divided by the wheel radius. Unfortunately, the size of this number is not very useful because we would need to compare it with the rolling and aero drag to decide just how much was left over for acceleration. That gets us into the realms of dynamic simulation and, while that is definitely the way of the future, we have the more immediate problem of making the car go better at its next event.

The shape of the torque curve is important because it will tell us how the engine delivers its power. If the torque curve is actually flat – impossible in reality – then the power will increase in proportion to rpm, so the lowest possible gearing that will reach the speeds that you need is the way to go. The lower the gearing, the bigger the shove and the better the acceleration.

Sadly, our engine will run out of air and rpm at some point and we'll need to change up. The best point to do this can be found either mathematically or by looking at the data. The mathematical approach needs a spreadsheet, a power curve, and the ratios. What you need to do is to create a table showing torque at 500rpm intervals, and multiply those figures by the gear ratios in each gear and the final drive and then divide by the wheel radius. This gives the thrust at the driving wheels. This data can then be used to plot a curve against speed that shows the thrust available at the contact patch. Do this for each ratio and plot the curves on one graph and you have the definitive document on when to change gear. By now the spreadsheet users will be beginning to see how they can do this for themselves, and the non-believers will be thinking about skipping to the next section.

Chart 8.01 shows this kind of plot and, even if you have no intention of preparing your own spreadsheet, there are some lessons.

1. The shape of the thrust curves depends absolutely on the shape of the torque curve
2. The lower the gearing, the higher the thrust and, therefore, the acceleration
3. The optimal shift point is where two curves intersect
4. The bigger the gap between two gears, the higher the speed at which the shift should take place.
5. Drivers can't carry 3 or 4 optimal shift points in their heads, so you need to compromise by rounding the numbers.

So, a starting point is the engine's power or torque curve. Some logger

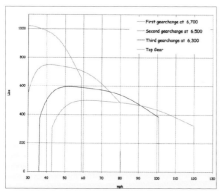

8.01. A plot of thrust at the tyre against road speed. The intersections of the lines show the optimal shift points.

software offers you the option of calculating a power and torque curve for the engine. This is only marginally useful because it depends on good experimental conditions, with no wind and a level track. The calculation needs you to supply the frontal area and the coefficient of drag, neither of which is usually available. Taking all these things into account, and bearing in mind how small the margins are that we're trying to work on, any logger-produced power curve can only be regarded as interesting rather than conclusive. One area where it can be of some use, though, is in giving a broad view of the shape of the power curve – helping you to see whether there are any dips in the delivery of power. This is particularly the case with kart engines, where the special nature of the power curve of highly-tuned small engines deserves special consideration

With something more conventional, selection of the right point at which to change gear becomes a matter of looking through the data to compare the effects of different rpm limits. Chapter 12 deals with this by suggesting that a good practical way round the problem is to use the early laps of qualifying to try gearchanges at different engine speeds. The data can then be analysed to see which rpm was best.

If it's possible to possible to change the gearing on your car, it is something you should bear in mind when preparing for an event. It doesn't matter whether the choice of ratio is limited just to the final drive or whether it's possible to change the individual gear ratios, logged data can be used to evaluate the best choices.

Teams running modified production cars are likely to carry different final drive ratios for different tracks. Motorcycle-engined cars achieve the same effect by having a choice of sprockets. With many purpose-built motorsport transmissions, there will be choices for the individual intermediate gear ratios. If these can be swapped quickly, there's more choice and more complication. When only the final drive is changeable, the target is normally to maximise the revs at the end of the fastest straight. However, when all the ratios are optional, we need to provide for intermediates that will carry the car through the important corners. Just occasionally, a team may choose to

sacrifice outright top speed for better acceleration on the rest of the course.

DISPLAYING DATA RELATING TO GEARING

The traditional way of showing gearing is by the use of a chart that shows engine revs and road speeds. Car magazines use them in their road test data panels, and competition gearbox manufacturers show the range of available ratios in this way. They are calculated with reference to the intermediate and final drive ratios and wheel size. A logging system can draw one quickly and easily simply by providing an XY graph where road speed is on the X axis and engine speed on the Y axis (see the accompanying chart).

Being able to choose what gearing to use means that we need to keep proper records so that we know exactly what setup was being used the last time at this track. A chart like this does the job for us. It's an instant, no cost, no argument record. Not only that, it takes into account not just the actual gear ratios themselves but also all the other things that influence the gearing, such as transfer gears (if fitted), final drive, and tyre size. A note of caution here is that this is only true so long as you measure speed correctly. A GPS system does this automatically, but if you're in the habit of changing tyre sizes, speed will need re-calibrating each time.

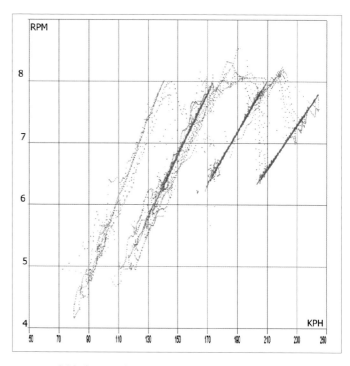

8.02. A gear chart created direct from the logger.

GEAR NUMBERS

Some systems log the actual gear that is selected, either by direct use of a sensor on the car (quite popular with bike-engined cars) or by some software-based solution that requires you to enter the ratios into the program. The software solutions use a programming algorithm to deduce what gear the car is in, and these are not always successful, but if you have access to either channel it's useful data that should not be overlooked. Chart 8.03 shows what it would look like as a strip chart, and later in this Chapter we look at presenting the gear information in map form.

Miles per hour per 1000rpm

The problem is that neither the software nor hardware solution to the gear problem is absolutely bullet-proof, which is why mph/1000rpm or a gearing ratio channel is so useful. This provides better information because it:

• Shows unequivocally what gearing was in use on the day
• Indicates the ratio selected
• Can highlight traction issues
• Is an indicator of poor driving technique

The maths channel is a very simple calculation of:

Mph per 1000 = speed ÷ (rpm ÷ 1000)

and it also gives an instant measure of the actual gearing used. It's a good idea to prepare a spreadsheet showing the range of gearing choices available to you – it will be helpful when you're making decisions about the choice of ratios and will add to your understanding of the data. If we assume that the gearbox ratios are fixed, but that there's some choice of tyre size and final drive, the chart could like Chart 8.04

In this table, the gear ratio is the ratio of each intermediate gear and is obtained by dividing the number of teeth on the driven gear by the number of teeth on the driving gear. In this way, a first gear where a 15-tooth gear drives a 40-tooth gear, the ratio is 40 ÷ 15 = 3.2:1. The intermediate ratios and the final drive are reduction ratios, so, with a 3.9 final drive, the overall reduction will be 3.9 x 3.2 = 12.48. This is shown in the third column as overall gearing

To calculate mph/1000, we need to take into account the fact that for every 1000 engine revolutions, the driven wheels revolve by 1000 ÷ Overall gearing. So, with an overall gearing of 12.5:1, they will rotate 80 times (1000 : 12.5). This is then multiplied by the circumference of the tyre (in units of your choice), multiplied by 60 to convert the minutes to hours, and divided by a number that converts the measurement of the circumference into miles. In the table, the circumference was measured in millimetres, so this

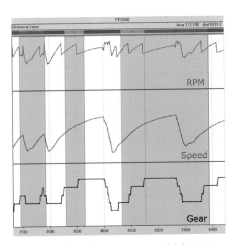

8.03. A strip chart showing which gear is selected based on the manufacturer's software solution.

Gear ratio		Final drive ratios		Wheel circumference	
3.2		3.9		Dry	1.6m
1.916		3.7		Wet	1.75m
1.357		3.5			
1					
Max rpm	7500				

		Overall gearing	mph/1000		Max speed
FD	Gear	Dry	Dry	Wet	Dry
3.9	1	12.5	4.8	5.3	36.4
	2	7.5	8.1	8.9	60.7
	3	5.3	11.4	12.5	85.8
	4	3.9	15.5	17.0	116.4
3.7	1	11.8	5.1	5.6	38.3
	2	7.1	8.5	9.3	64.0
	3	5.0	12.1	13.2	90.4
	4	3.7	16.4	17.9	122.7
3.5	1	11.2	5.4	5.9	40.5
	2	6.7	9.0	9.9	67.7
	3	4.7	12.7	13.9	95.6
	4	3.5	17.3	18.9	129.7

8.04. A ratio chart compiled in Excel for a car with fixed gearbox ratios and a choice of final drives.

Tooth count	mph per 1000 rpm	Speed at Mmx revs of 6500
25/25	21.61	140.5
25/26	20.78	135.1
25/27	20.01	130.1
24/26	19.95	129.7
24/27	19.21	124.9
24/28	18.52	120.4
23/27	18.41	119.7
23/28	17.75	115.4
23/29	17.14	111.4
22/28	16.98	110.4
22/29	16.40	106.6
22/30	15.85	103.0
21/29	15.65	101.7
21/30	15.13	98.3
20/30	14.41	93.7
20/32	13.51	87.8
19/32	12.83	83.4
19/33	12.44	80.9
18/32	12.16	79.0
18/34	11.44	74.4
15/35	9.26	60.2
15/36	9.01	58.5
14/36	8.40	54.6

8.05. This ratio chart goes to the track with the car. Logging rpm/1000 allows the team to see what ratios were fitted in any session.

will need to be divided by 1000 to convert to metres, and then by 1000 to convert to kilometres, and then by 1.61 to convert to miles. If you measure in feet, this needs dividing by 5280 for the conversion. The max speed is the mph/1000 multiplied by the max rpm.

With a competition gearbox, the manufacturer is likely to publish the speed per 1000rpm figures for each of the available ratios using various tyre sizes, and this saves you the (pretty minimal) effort of doing your own calculations. If you feel the need to do your own gear chart, something like the one shown in Chart 8.05 will do the job. Here, the assumption is that the choice of intermediate ratios is sufficient so that you don't need to alter the final drive ratio or tyre sizes.

In this table, the gearing ratio is the numerical ratio obtained by dividing the tooth count on the driven gear by the tooth count on the driving gear, and this is multiplied through the maximum rpm and the wheel circumference to calculate the speed and the formula for speed looks like this

Rpm ÷ gearing ratio ÷ final drive x wheel circumference in miles x 60

This is as accurate as you are in measuring the wheel circumference, but gives a figure that can be compared with the data taken directly from the logger.

So, by simply reading the mph/1000 figure off the logger and comparing it with the gear chart, you can read off what ratio was used for that outing. The error is usually only a couple of tenths of a mile per hour, so identification is pretty reliable. Chart 8.06 shows what an mph/1000rpm trace looks like.

Maths channels were introduced in Chapter 5 and one of the examples used was the mph/1000. Refer back to Chart 5.10 to see how it was calculated. The strip chart here in Figure 8.06 shows mph/1000 plotted against distance. You can see that the trace is mainly made up of straight(ish) horizontal lines. These indicate the gear ratios and show 12, 15, 18 and 21.5 miles per hour per thousand rpm. Reading off the gear chart (Figure 8.05) this means that the ratios in use were 18/32, 21/30, 23/28 and 25/25. Knowing that the figure of 21.5 relates to fifth gear, we can deduce that first gear is not being used during the lap. The transitions from one horizontal portion to the next represent gear changes up and down and the nearer these are to vertical, the quicker the change is happening. Spikes upwards represent poor technique, where the wheel drags the engine back up to speed and spikes downwards show lack of traction.

A maths channel to identify the gear number

If mph/1000 does not appeal to you,

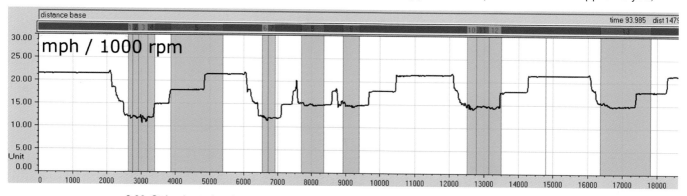

8.06. Strip chart showing mph/1000rpm. Each step up or down represents a gearchange.

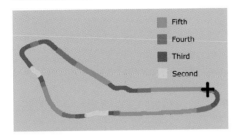

8.07. This is a map report, coloured according to the gear that is selected.

and you absolutely must have a channel showing which gear is being used, then there are maths solutions which will do the job, but they are complex and require multiple 'if' statements. They work by starting with the number 1 and adding in a further 1 every time the mph/1000 value is more than the midpoint between two ratios. To demonstrate the technique, let's consider the case of a four-speed box with interchangeable ratios, with first gear going up to 14mph/1000, second gear up to 18, third up to 22 and top 26mph/1000; the midpoints would be 16, 20 and 24. The formula would then be:

Gear number = 1 + (if the mph/1000 >16, add 1) + (if the mph/1000 >20, add 1) + (if the mph/1000 >24, add 1)

This then needs to be translated

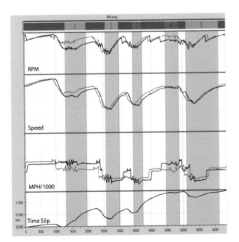

8.08. Minor changes to the ratios provide faster lap times.

into the maths language of your system. Done correctly, it will return a trace that gives nice crisp straight lines indicating what gear is being used. It can get confused if you're using oddball gear sets with two gears very close together, and it loses an awful lot of information that is available in the straightforward mph/1000 trace.

The gear map

One very good application of the data-coloured maps that most software provides is to show mph/1000 as a map (shown at 8.07). The colours show which gear is being used and this is a great reminder of how to drive a particular track (a copy can even be taped to the steering wheel).

We can quickly work out by looking at the data or talking to the driver that fifth gear (green) is being used for the straights, fourth (blue) is a straightforward intermediate gear, fast corners are taken in third (red), and the chicanes in second (orange). The map does show up a problem. We can see that the driver is selecting fourth gear briefly at the top of the circuit and probably a taller third gear would save two gear-changes and up to half a second.

USING THE DATA FOR DECISIONS

The gear map has an immediate impact, but for proper analysis nothing beats a strip chart showing mph/1000 and other channels. Looking at the trace on its own shows the actual gearing used. Used with rpm it gives a view of where the engine is running out of revs or is having to operate below a comfortable threshold.

Chart 8.08 shows two laps from a test day during which the ratios were changed. The new ratios relate to the red trace and involved using a slightly higher top gear for better top speed, and higher

third gear to avoid an unnecessary gearchange at one point of the lap and a higher second gear to reduce the 2nd/3rd gap and improve acceleration. The cost of this was lower rpm in Turn 2. The table shows the ratios before (black), and after (red).

The black time slip line climbs when the black lap is losing time. It shows the major benefits of the new ratio set. Turn 1 can now be taken in third rather than top gear, and allows better control and better speed. The higher second gear gives a better punch out of Turn 3 and saving a gearchange between Turns 2 and 3 presses home the advantage.

Choosing the optimal ratios

Sadly, there's no foolproof way of optimising the gear ratios, but there are some things that can be done to improve your chances of getting it right. When going to a new circuit, a good starting point is to take advice from someone with experience there. Be careful, though; competition being what it is, you're not likely to get the optimal set from someone who thinks that you might beat them. Another point is that if you 'borrow' ratios from the front runners and you're still in the mid-field, then their ratios are likely to be a bit high for you simply because their corner speeds are going to be greater, and this will translate into higher speeds on the straights. So, knock a few percentage points off any set you find in this way.

Where intermediate ratios can be changed, the job is to provide a suitable top gear, a first gear that gets the car off the line competitively, and intermediate gears that fill the gap nicely. Generally, the higher gears need to be closer together than the lower gears to cope with the effects of drag at high speeds.

The gear chart shown at 8.09 was drawn using Excel rather than the logging software and shows what you are trying to achieve. The reduction in

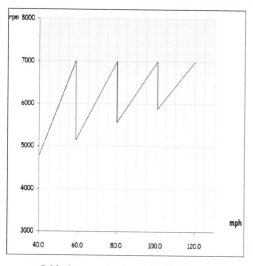

8.09. A gear chart created in Excel.

rpm at each gear change gets smaller as the speed increases. The revs fall from 7000rpm to 5150 when changing from first to second gear, 5550 on the change from second to third, and to 5900 from third to top.

The problem with this theoretically correct approach is that it doesn't tell you how much closer the gears need to be, and it doesn't deal with gradients or help with the important corners. At most tracks, speed through one or two bends has a big impact on the lap time. This may be because a long time is spent in that particular bend or the bend feeds on to an important straight and any speed lost at the start of the straight hurts all the way to the next corner. In the UK, Gerrards Bend at Mallory Park takes up about 12 seconds of a 55 second lap, and the Revett Straight at Snetterton takes 16 seconds of a 75 second lap. In these and other cases it well might be necessary to abandon the theoretically perfect, choose a ratio for the important corner, and limit the damage elsewhere.

If the choice is limited to final drive ratio (i.e. it's not possible to change the ratios in the gearbox), then things are relatively simple. We should be looking for a final drive that allows top gear to exceed the rev limit at the fastest part

of the course by a non-damaging number of rpm. You should always be well past the point of peak power because you are trading losses at the top end against the benefits of lower gearing. The truth is that if you work out the available torque as the car approaches top speed and subtract the aero and rolling drag, you're left with a very small number of Newtons or pounds of thrust to accelerate the car. So, at the top end, even though it may be well beyond peak power, having the benefit of an extra bit of torque multiplication from a lower overall gearing pays unexpected dividends.

Luckily, this is exactly the sort of question that you can answer with certainty with data. Chart 8.10 shows gearing with about 5 per cent gap (23/27 as opposed to 24/27). In this case, the peak rpm was 6700 as against 6200, with peak power at around 6300. Speed difference was 4mph, and the time saved 0.31 seconds just for the top gear blast. These are the sort of areas where data logging can pay huge dividends.

Having dealt with the overall gearing, it's now time to turn to deciding whether the gearing is the best possible. Looking at Chart 8.11, point 1 is included just for interest and indicates that the driver selects each gear in turn under braking, rather than going from the current to the required gear in one step. This is inevitable if the box is sequential, but if it's merely a matter of preference, it might be worth some further thought. Point 2 on the strip chart shows the driver going up through the gears and, other things being equal, you would expect to see a more consistent cascade of steps rather than the uneven spread indicated by this collection. Point 3 indicates a brief burst of fourth gear

8.10. The highlighted areas show the effects of exceeding peak power in top gear.

before two corners. Zooming in and measuring shows that the gear was only engaged for about 80 metres, and the possibility of a higher gear has already been discussed. Point 4 shows that each time the driver changes down from top gear, this is done a few hundred rpm short of the rev limit, so there may well be some benefits from lowering the overall ratio.

Looking at the corners in the chart, it's apparent that second gear is well chosen for the first complex of corners, since it just reaches the rev limit at the exit. There is a compromise, in that the rpm does fall to quite a low level in this corner, so lowering second gear might be desirable, but if so it would need a gear change sometime during the series of bends. The driver's opinion should be asked. The alternative would to be to use 1st gear to keep the engine in the meatier part of the power band, and again an upshift would be necessary. Changing up in Turn 5 (the second shaded column) obviously presents no problem. The second chicane (the third

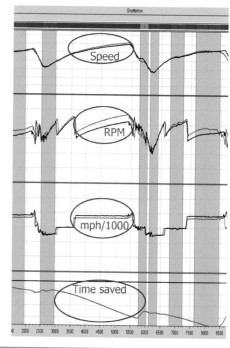

shaded column) also falls off the power curve, so there might be some mileage in the high first gear strategy here also.

Lap times should tell us the way to go, but the data is a real help. It's sometimes the case that you can devise a set of ratios that does not find any time per se, but the driver feels more comfortable with it. The worst thing that can happen is for a beginner to copy blindly the ratios of the front runners who are likely to be a significant amount faster at all points on the track. Choose ratios realistically to suit your current level of performance. When you get closer to the front runners, then you can think about copying what they do.

Traction

It is possible to create a maths channel that measures percentage wheelspin, but this is complex and needs a series of 'if' statements to identify what gear the car is in so that it can eliminate gearing effects. If you feel that you must, go ahead, but the mph/1000 is a pretty adequate substitute.

In Chart 8.12 the lowest gear used provides a very noisy trace (area 1 of the chart, magnified just below it) and this is indicative of a mild level of wheelspin and suggests the car is not fitted with a limited slip differential and is using the kerbs. It could have been much worse, but it is always worth investigating.

Upward spikes show that the engine revs are dropping in relation to the road speed, and are indicative of the driver being careless with the clutch because the gear is selected and the clutch released clumsily, forcing the engine revs to catch up with a jerk. This is the sort of time when you need to use the data and the driver together to discover the whole story. These are shown with the legend 'gearchange technique' on the strip chart.

If the spike in mph/1000 is downward, revs are higher than

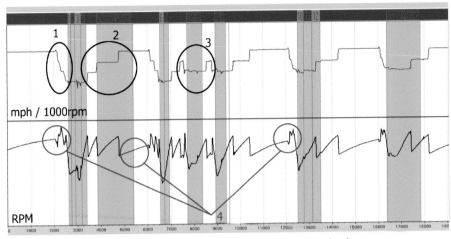

8.11. Using the mph/1000rpm chart to sort out gearing issues.

expected in relation to the road speed. The simple explanation can be either dipping the clutch or momentary wheelspin. Not good for either the drive train or the grip at the rear wheel, and time for a word with the driver.

A 'noisy' trace with spikes in both directions through the corner is symptomatic of either traction problems or the car being driven absolutely on the limit. With an open diff, hitting the kerb can launch a wheel and cause the revs to rise, and the data can be used to point out to the driver that time spent spinning the wheels is time that could otherwise be used for acceleration. Mid-corner, away from the kerbs, the noise suggests lack of traction due to a needlessly lively suspension.

Traction can be a real issue with some cars, and the logger provides us with some insight. Mph/1000 is a good starting point, because any wheelspin will have the trace heading downwards because the speed stays pretty static while the rpm climbs. This is much more easily seen than it would be on the rpm trace. With severe wheelspin, the rpm trace will spike upwards, but mild wheelspin simply tends to show up as a slight hump in the curve. Look at circled areas in the top half of Chart 8.13 and you can see how much easier wheelspin

is to spot on the mph/1000 trace than on the rpm trace.

The bottom half of the chart shows a low-powered spec racer where wheelspin is definitely not a problem.

Differentials

Shown in conjunction with other handling channels (lateral and long g, steering and throttle), the logger can tell us what the limited slip diff is doing. All of the traction pointers discussed above might appear, but in much more muted form. With limited slip diffs there will also be effects as the car turns into and exits the corner, and it's this that we need to be aware of. The plate type (or clutch pack) diff starts with an amount of preload that splits the torque equally up to a specified amount. So, while this is happening, the driving wheels are going to give you no assistance in turning into the corner. They're going to continue to push the car forward in a straight line. So, this is data clue number 1 – if the car suffers initial difficulties in turning in this might not only be attributed to the damper settings, but might be a diff pre-load issue. The channels to look at are steering and throttle.

Once the initial preload is overcome (usually this shouldn't be an issue) the braking effort on each wheel loads up

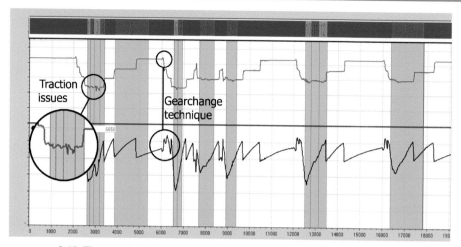

8.12. The mph/1000rpm trace can show up traction and driving issues.

the plates and splits the torque between the wheels according to how much

each can handle. This stage should be completely under control because the

lsd is a device for sending the torque where it can be best used. But things can go wrong. The diff may not work as well as it should, and this may lead to wheelspin or equal sharing of the torque which will hinder the change of direction. Some plate-type differentials use ramps to modify the rate at which the slip limiting action takes place. A steep ramp angle means that the lsd kicks in rapidly, a shallow angle makes the effect more gentle. So, this is the next data clue; what does the data tell you about phase two of this operation – the transition from turn in to mid-phase of the corner? Look for sudden and unpleasant changes to rpm and driven wheel speeds (if you log them) that don't seem to occur as a result of a clumsy driver. On the way through and out of the corner, the reverse happens. The preload dictates the point at which the ramps kick in. The rate at which they do is dictated by the ramp angle. This side of the Atlantic with our punier engines, 30lb of preload is considered mild and ramp angles of 45° are considered reasonable. If you decide to change the ramp angles of the diff, be sure that you understand how the manufacturer specifies the angle and what is considered a suitable range from mild to fierce so that you can make your choice accordingly. Don't forget that you should be able to specify different ramps for corner entry and exit so that you can influence the turn in and the power application parts of the corner.

In all cases, the logger will provide you with the information you need to make good decisions. Track testing is the only way to be certain, but intelligent use of the data will reduce the number of ratios that you need to try.

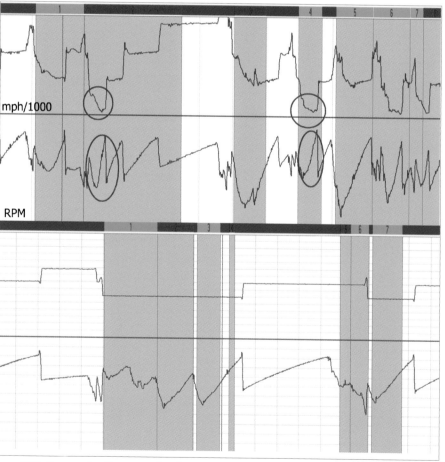

8.13. Two contrasting cars. One has a powerful V8 and the other is a four-cylinder production racer: the mph/1000rpm trace shows distinctly different traction patterns.

Chapter 9
Brakes

Brakes are definitely optional – from the logging, not the racing viewpoint. You can collect braking data in several ways.

• Brake light switch
• Position sensor on the pedal
• Pressure sensor (in one or both circuits)
• Mathematical modelling of braking effort
• Interpreting the long g trace

The first three options are hardware related and involve modification to the braking system, so you had better be sure that you don't interfere with its safe and efficient operation. The hardware options measure the timing and/or ferocity of the input (the use of the driver's right leg). The other approaches use the software to evaluate the output of the system and give some idea of the rate at which the car is slowing down. In a perfect world, the team would have access to both sets of data, but either will help us stop the car more quickly and get it into the turn better.

BRAKE PEDAL SWITCH OR MOVEMENT SENSOR

If rules require you to have operating brake lights, then it makes sense to use whatever mechanism switches the lights to send a signal to the logger. There might be compatibility problems here. If you use an analogue channel, then the voltage needs to be compatible with what the logger expects. If this is 12V or the logger can cope with that sort of over-supply on a 5V channel (check carefully first!), then the switch will provide a signal that will show as either on or off. At least it gives you some view of the timing of the braking, and the

similarity with the single line pressure traces (Chart 9.02) is obvious, the only difference is that a switch shows the timing, not the amount. It also records when the brakes were applied and released. With access to several laps' worth of data, it's possible to evaluate the costs or benefits of late braking.

It's possible to log the position of the brake pedal with a potential divider, but it's not a very common option. It will show how far the pedal travels, and this will give some indication of the extent to which the brakes fade and the pedal gets 'long,' but, on the other hand, most drivers are quite anxious to

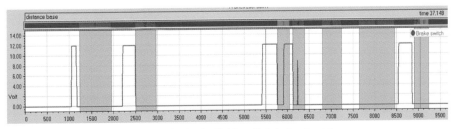

9.01. Using the brake light switch to log braking.

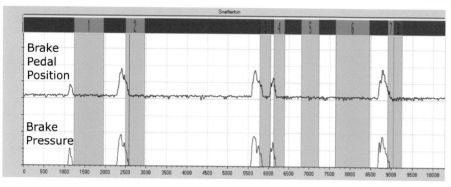

9.02. Logging the position of the pedal is a useful way of looking at the braking, and mimics a pressure trace quite effectively.

to 0.3sec. Anything longer than this and the driver is getting out of the throttle early and taking a small confidence wait before the brakes are applied. Time for an exchange of views with him or her. If the gap is 0.3sec, there's the possibility that the cockpit layout is not as required and, although the driver is braking late, the mechanical linkages are just not getting the job done. None of this should apply to drivers who use left foot braking. In this case there should be no gap at all between deciding to end

tell the team about a long pedal. It's a surprisingly good substitute for logging a single line pressure, as can be seen from Chart 9.02. When there are several laps available for comparison, it's possible to evaluate the benefits or costs of late or early braking.

If you do any form of brake logging, it's useful to create a maths channel that shows the integral of one of the channels that you log. This will give you the total time that you spend on the brakes during a lap. Comparing session to session would show how effective the brakes are, and comparing braking time to lap time gives an idea of how hard on brakes a particular track is. Chapter 5 shows a channel report for braking data in Figure 5.06.

PRESSURE SENSORS
Pressure sensors that will operate in the pressure range required to measure the brakes' hydraulic pressure are expensive, and if you're unfortunate to choose the wrong one, fragile. So find a professional team and talk to their data engineer to find the name of a reliable brand before parting with your cash. Ideally, you need one sensor in each circuit. Fitting one to each wheel doubles the cost and gives you very little extra information.

If you can only afford one sensor or one channel of your logger to record

brakes, there is still useful information to be had. A pressure sensor will show the point at which the driver got on to the brakes, the rate at which they were applied, and the way in which they were released. The trace can also be related to the long g trace and the combined g trace to give some understanding of how well the driver is carrying out the braking part of the job. An example of a pressure trace can be seen in the lower half of Chart 9.02.

The first thing to look for is the time taken to apply the brakes, so the chart needs to be plotted against time, not distance. You can identify when the driver gets off the throttle and measure the time that it takes for the brake pressure to build up. The increase in pressure should be rapid. A good driver takes about 0.2sec, and an average one up

9.03. Braking with aero downforce – a high pedal pressure reaches maximum braking effort but, as the downforce goes away, the driver has to reduce the pressure to avoid locking the wheels.

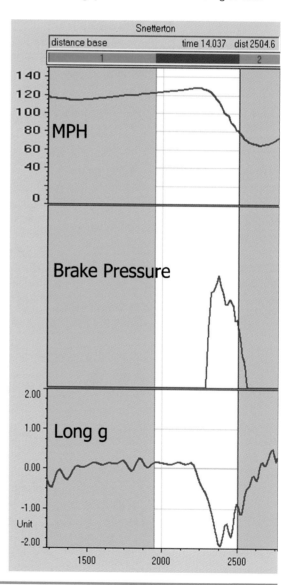

accelerating and to get on the brakes. In fact, most drivers in this situation tend to just brush the pedal to take up the slack in the system and to build up a bit of heat in the pads immediately before getting out of the throttle.

The shape of the pressure trace will vary according to the type of car involved. A car with downforce should show a very high pressure immediately and then show the driver backing off the pedal as the speed comes down and the downforce is no longer available to support high levels of braking. A hatchback with a high centre of gravity and a short wheelbase needs a much less aggressive curve, with the driver taking care not to transfer weight forward too dramatically at the start of the braking and probably coming off the brakes equally gently to avoid upsetting the car during the turning in phase. Yet another instance of not treating the data in isolation but involving engineer and driver together.

TWIN SENSORS

Two sensors will also provide a measure of when and how hard the brakes have been applied, and also enable you to calculate the brake balance. In addition,

they can isolate some otherwise very hard to find handling problems.

Calculating the brake balance by using the actual line pressures does not tell you the real balance and the amount of work done by the brakes at either end of the car. That would depend on the sizing of the various components and the amount of weight transfer during heavy braking. What it will do is give you a rough and ready indication of how the balance was set, and allow you to compare the effectiveness of set-up at different events. It also gives you a good diagnostic tool.

Although it's not essential for the pressures to be equal, it's good practice to keep them broadly in line with each other. Failure to do so can lead to inconsistencies in braking, as the balance bar is used to provide the right balance, and this is only successful if the balance bar remains reasonably square during operation.

So, by logging line pressures, we can see that the pressures are broadly similar, and a maths channel will give us a number that we can call the 'brake balance'. This is not the balance of actual braking effort, but the relative pressures, which gives us a measure of how the balance was set up. And, by comparing it to long g, we can see what line pressures (and therefore balance) work best in the wet and the dry. If we're after really fine tuning, we might even try to distinguish between grippy and slippy track surfaces.

Any equation that relates the two pressures will do, but the data shown in this book is based on:

9.04. Twin brake pressure sensors, plumbed into the brake lines using tee pieces and braided hoses, and provided with bleed screws.

Front pressure ÷ (front pressure + rear pressure) x 100

This gives a percentage showing the front brake pressure as a percentage of the total. Greater than 50 would indicate that the front pressure was greater than the rear, and vice versa.

There is often a problem here with spurious readings. If we calculate the ratio between front and rear circuits there will be occasions when we get nonsensical results. For example, when the car hits a bump or a kerb, the pedal might rattle and put a very small amount of pressure into the system. If it's only the front system that is disturbed, the ratio calculated will be a very large number and will show up as a spike in the data that makes the trace look messy.

There are two solutions. One is to calibrate the system so that it only starts reading at some arbitrary low number (say 0.5bar or 5psi). The other is to filter the noise out of the system at the maths stage by using a high-pass filter on each pressure channel, and Chapter 4 tells you how to use a high-pass filter to select only the values above our minimum.

Chart 9.05 shows the two line pressures, and the overall balance of the brakes. The balance trace is the shape that you would expect to see, with front and rear pressures changing together and keeping the balance figure nice and steady. The key here is not so much the numerical value as the shape. It rises rapidly as both circuits increase in pressure, is consistent throughout the braking event, and falls rapidly when the brakes are taken off. This is simply a representation of the balance.

The effects of a bump rattling the brake pedal are highlighted. The line pressures generally show that the driver is not totally comfortable with the brakes, since there are many instances of the

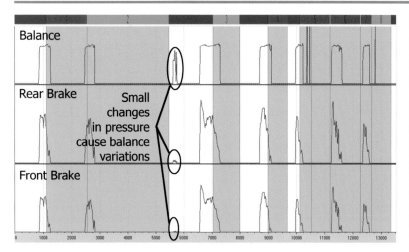

9.05. Brake balance – the block-shaped trace indicates that everything is operating properly.

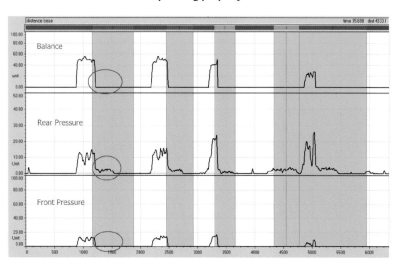

9.06. Rear brake staying on, not showing up in the balance calculation.

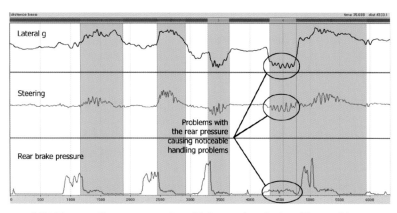

9.07. The rear line pressure stays high, causing the handling problems indicated by the lateral g and steering traces.

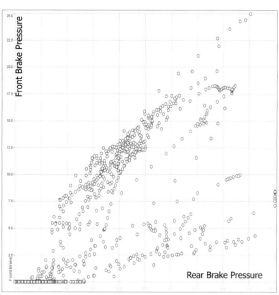

9.08. The braking problem shown in Chart 9.06 shows up clearly in an XY chart.

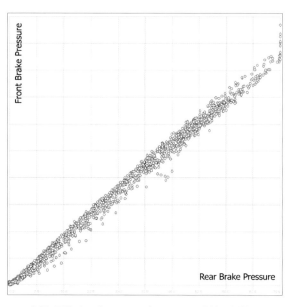

9.09. This is what a good trace would look like.

pressures being built up and then released during the stop.

If the brake balance trace does not have this flat, brick-like appearance, it's a warning sign. If the balance climbs slowly or peaks suddenly at the beginning of the braking effort, it's indicative that one brake is being applied much more rapidly than the other, so it's worthwhile looking at the balance bar and linkages to see why. The same causes might be responsible for a peak at the end of the trace, or a sloping down rather than a sharp cut off.

The end of the trace symptoms are the worrying ones. If the rear brakes are not being released as quickly as the front brakes, and the driver turns in under braking, then only the rear wheels are being braked during the turn-in phase, and this can create oversteer that the conventional cures won't touch. The problem is not with the springs, dampers or geometry, but with a mechanical malfunction of the brakes.

Unfortunately, the balance trace in Chart 9.06 does not show the rear pressure failing to release, because the front pressure is zero and, therefore, the balance percentage calculation returns zero. An instance of this is shown highlighted in red. The handling did cause the driver to work hard, though, as can be seen in Chart 9.07.

It was a pity that the brake balance trace showed no evidence of the problem, but an XY chart plotting front pressure against rear pressure showed the position quite clearly. In 9.08 the data is very scattered, which indicates a problem, and should prompt an investigation into the mechanical action of the balance bar system. The tight grouping of the data in 9.09 is what you would expect to find, and indicates that all is well.

LONG G FOR BRAKING

The simple long g trace shows how well the driver is controlling the brakes and how well the car responds. The good point about this is that it shows the output (how quickly the car is slowing), rather than the input (the way in which the driver is applying the brakes). We get some clues, but more would be better.

In a perfect world, we would have enough tyre data to know the absolute maximum braking effort that the car

could achieve. The tyre companies don't provide it, so we need another strategy, and about the only one that works is trying to work out how well our competitors are performing.

This is another case where sharing data can really pay off, so if you can find someone who will swap data with you, it's well worthwhile. Without access to other people's data, the driver should still have a fair idea of how effective the car is under-braking. The driver will spend a significant proportion of the time in company with other cars, so he or she should know how the brakes compare with the opposition without the benefit of logged data. However, it's an opportunity to get things wrong because most drivers believe that the brakes are excellent.

The evidence is there to be seen. Along the straight, the gap between our driver and the car in front might be 50 metres. At the apex of the following hairpin, the gap might be down to 15 metres. Therefore, the brakes are good enough to pull back 35 metres in one go. The flaw in this logic is that gaps are best measured in terms of time. At the end of the straight, the cars were probably travelling at 50 metres per second. At the apex of a tight bend the speed might have fallen to 15 metres per second. So the gap between the cars was 1 second at the end of the straight and the same at the apex of the

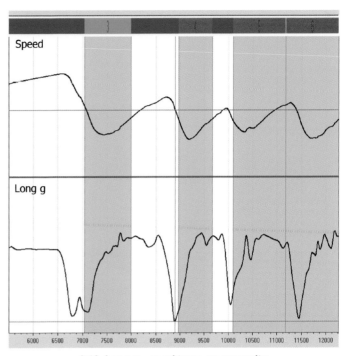

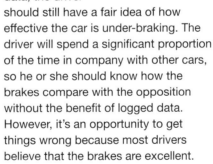

9.10. Long g – good trace, poor results.

bend. The braking performance must have been identical, although it certainly doesn't look like it to the driver, who, incidentally, will tend to think that other cars have more power and pull much better out of corners, by using exactly the same logic. With a properly briefed driver, we should be in a position to understand the long g trace.

Chart 9.10 shows four distinct braking areas, and is what you would expect to find with a racing car fitted with adequate brakes. Whenever the trace is below the zero line, the car is slowing down, so even as the trace climbs back up towards zero, the brakes are still on, but being released. So, for a driver who trail brakes, we would expect to see a steep plunge when the brakes were applied rapidly, and then a more gentle climb back up to zero as the brakes were released on turn-in. This trace is fairly 'peaky'. It reaches the peak value rapidly and then reduces almost immediately. The brakes are adequate, and used well. There is one problem, however. The peak long g is not consistent. Long g is better before Turns 4 and 6, and poor before Turns 3 and 5. This is not an aerodynamic effect, because the highest values are on medium speed corners. The first braking area also shows evidence of braking too early because the driver backs out of the pedal a little bit before getting back on. These are harsh criticisms. Amateurs generally do not use their brakes anything like as effectively as professional drivers.

A characteristic of shortage of brakes is that the long g trace falls to its maximum value and the driver spends some time at this value before releasing them. Chart 9.11 shows this clearly. What we can deduce from this chart is that if the brakes are inadequate, the trace is longer and shallower. This should give us some clues about how to recognise brake fade. If, over the course of a long run, the shape of the trace varies, and moves from spikes into longer, shallower blocks, then we should suspect brake fade, and investigate.

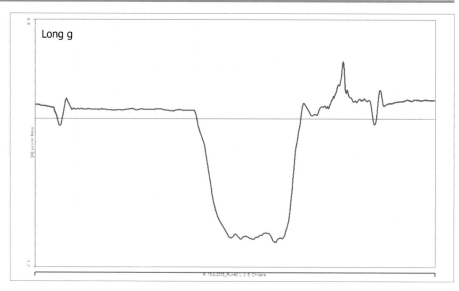

9.11. Braking trace from production-based car – note the significant period of time spent at maximum effort.

BRAKING EFFORT

The problem with the long g trace is that the car will slow down naturally whenever the throttle is lifted. It would be nice to isolate the effects of the driver using the brakes from this naturally occurring loss of speed. This can be done using a coast down test and a bit of maths. We can calculate the braking effort by subtracting the drag from the long g.

To do this, we need to know the level of drag, and this can be found using coast-down testing. At some convenient point the driver de-clutches and lifts off the throttle. The car is allowed to lose speed for a reasonable amount of time and this provides us with a negative long g trace. The next steps are much less complicated than they sound.

First, we must export the data to Excel and create a chart showing the deceleration. We need an XY chart where speed is shown on the X axis and long g on the Y axis. If we then add a trend line to the chart we can display the formula of the trend line on the graph. This formula will be something like:

$$y = 5E\text{-}08x^4 - 1E\text{-}05x^3 + 0.0014x^2 - 0.0668x + 0.7903$$

where y is drag expressed in g and x is the speed. This method is described more fully in Appendix 1. It makes it possible to create a maths channel that will calculate the drag at any speed as:

$$\text{Drag} = 5E\text{-}08\text{Speed}^4 - 1E\text{-}05\text{Speed}^3 + 0.0014\text{Speed}^2 - 0.0668\text{Speed} + 0.7903$$

and

$$\text{Braking effort} = \text{long g} - \text{drag}$$

This method is a bit long-winded, but it mimics a pressure channel quite well and provides a surprisingly useful amount of information. It's a useful solution for a team where the money for pressure sensors could be better spent elsewhere.

A proper scientific coast-down test would involve a flat track, hot tyres, and runs in opposite directions to minimise experimental error. The coast-down test that can be carried on at a circuit will always be subject to error so there is a case for doing them whenever the opportunity presents itself, and keeping all of the results to build up a bank of data that provides a useful average to work from. Spurious data can be identified and discarded. This approach is not available to cars with high downforce and high drag, because each aero setup will have its own level of drag, and frequent recalculations would be needed. Suddenly buying pressure sensors seems like a good idea.

Chapter 10
Running the system

After chapters on choosing and installing a system and how to interpret the data, all that remains now is the small matter of running the system successfully. Chapter 11 deals with the driver's point of view, and this chapter is written with the team manager/race engineer/data engineer in mind. This job will involve much more than simply reading the strip charts and other output, because without management of the logger, there won't be any output anyway. The sad fact is that it takes effort to run a data logger. It's easy for a professional team; they either appoint a data engineer or add it to someone's job description. For an amateur team where two or three people share all manner of tasks between them, things might get overlooked.

Someone has to take responsibility for the well-being of the system. The job starts back in the workshop as part of the routine for getting ready for an event, includes taking responsibility for its operation during an event, and checking it over afterwards. One of the most

effective pieces of equipment that you can have is an enthusiastic teenager with an interest in motorsport and computing. Ask around at the local high school or university. Ours came as a second year undergraduate, who went on to get a first class engineering degree and then a PhD. The PhD was a bit over the top, but it kept him in the neighbourhood.

PRE-EVENT WORK

Before every outing someone needs to do a physical inspection of the logger, the sensors and the wiring loom. If it's a dash logger, the functions of the dash should also be checked. The next stage is to check the calibration. When the system was installed, the set-up software provided a means of making sure that the sensors gave accurate data. The readings will tend to drift from their original settings, though, and it's best to re-calibrate the sensors than to fix the errors by adjustments in the analysis software.

For example, the simple O-ring

and pulley arrangement for measuring steering wheel movement can drift over time, and even a few tenths of a volt can cause steering traces from different outings to differ. This isn't life threatening, but it is irritating, and can be avoided with a bit of care. It's reasonably easy to use the analysis software to make an adjustment to the way in which the data is displayed, but it should not be necessary. Even if you use a toothed belt arrangement, there is the possibility that the belt may have jumped a tooth or two, so the only way to be certain is to hook the computer up to the car and look at the outputs in real time. You really ought to do this every time the car is set up. Professional teams will set up the car before each event and, although amateur teams tend to do this less frequently, whenever the car is on a flat patch presents the perfect opportunity to look at the calibration.

The computer can be plugged into the system and the online (or monitor) software used to check the following:

• All channels are providing a signal
• The signals read 0 in the appropriate place
• Full scale deflections reflect reality
• The lateral accelerometer is calibrated correctly
• Brake pressures are ok and reflect the balance specified by the engineers
• Wheel speed sensors record values of a few mph when the wheel is spun by hand

It's only a couple of minute's work to check that all channels are providing some sort of signal. The real time monitoring software will display all the channels on the screen, so quickly look down the column of boxes to ascertain that there are no zeros or values that will not move when provoked. Once that's done, you can move on to deciding whether the calibration is accurate.

The steering can be zeroed simply by setting the wheel in the straight ahead position, but it's better to have some physical means of centralizing the rack. Verify the logger zero and then look at full left and full right lock. This should be close to the maxima that you use in the analysis software. The accelerometers should be checked to see that they indicate zero when the car is on level ground, with driver and representative fuel loads aboard. Wheel speed sensors apparently take delight in failing without notice, so it makes sense to spin the wheel and check that there is an output. Check also that the throttle position sensor shows sensible readings. If they fail they still tend to give a signal, but it is highly inconsistent.

Most teams will run the engine to check systems prior to setting off for an event, so this gives the ideal opportunity to check the functions of the logger. It is particularly important that the rpm signal is clean and consistent, and you can also run a quick check on the measurement of temperatures and pressures. It is

better to log a couple of minute's worth of data rather than just using the monitor software. This is especially true with the rpm signal. It can look quite convincing when you simply blip the throttle and watch the numbers going up and down. The reality might be a bit different.

A good routine is to start the engine, let it idle for a while and then increase the rpm to 2000 and hold it there for three or four seconds, let it fall back to tick over and then increase the revs slowly and consistently to 3000, and hold it again. That is usually enough to test whether the signal is clean and accurate, and besides, the engine technician won't let you use any more revs no matter how good it sounds. The minute or two that this takes should cause the temperatures to climb slightly and allow you to look at them and the pressures as well. It's not checking that everything's working that takes the time, it's the fault finding if everything is not as it should be, and there'll be more time back at base than there is on an event.

Data logger fault finding

It can only be good news to have found a problem in the workshop rather than at the track. If your checks turn up a problem, using a logical approach will minimise the pain and the time taken to rectify it. You can make your life easier if you take the trouble to do a wiring schematic that shows what cables go where and which pins carry the supply voltage, the ground (0v), and the signal.

The logical place to start is the power supply. No sensor can return a signal unless it has the necessary voltage, so this is the first step. Check that the connectors into the back of the logger are properly seated. Sometimes, simply wriggling the plugs can make things work. If this fixes things, it should be regarded as only a temporary fix before you do whatever's necessary to prevent the problem recurring.

Something to watch out for here is that a connector pin might have got pushed back into the body of the connector, and a good wriggle makes a contact that fails again as soon as you turn your back. Check that the logger itself is providing the correct voltage at the output pin, and trace this voltage all the way to the sensor.

If the sensor has a suitable feed, check for a signal by connecting the meter between the signal wire and the logger ground and exciting the sensor. Set the meter to 0-20v for potential dividers or to the lowest AC voltage range for frequency generating devices like wheel speed sensors, and expect (hope) to see some movement on the meter. Seeing an AC voltage (or a frequency if your meter will measure that) is not foolproof, but it's the best that you will do without an oscilloscope. If you get no sign of life from the sensor, try a replacement. If the sensor works, then it's a matter of tracing its signal back to the logger unit and ensuring that it reaches the appropriate pin.

Connector blocks suffer from moisture much more than sealed connectors, although even these are not immune. In both cases the physical connection is in place, but an oxide film might be stopping the current. If this is the pre-event check, any water in the works will have dried out, but it may well have left corrosion that can prevent a good connection.

If the fault responds to these checks, but recurs intermittently, suspect a problem with the cabling. The easiest way out of this fix is to have provided a spare cable of each type at the installation stage. All of these instructions assume that the harness is such that you can get a meter into the pins, and with waterproof connectors that are pushed or screwed shut this is not always possible. Again, the counsel of perfection is to have bought or made some break-out type adaptors so that you can investigate the system. All that's needed is a plug or socket of each type, connected to a short length of cable so that you can check the voltages. Cables from old abandoned sensors are a good source.

If the accelerometers give unexpected reading you should investigate further. If the unit is placed on a level surface, it should read 0g. You can play about with packing shims to see how much is needed to get a reading of zero. Two or three degrees out is worrying, five or more is unacceptable. Small inaccuracies can be tolerated (and adjusted in the software) but big offsets prejudice the accuracy of the system. Refer to the manual or the manufacturer's website to see if there's a procedure for re-calibrating the accelerometers. If not, you'll need to return the system for it to be done.

AT THE TRACK

Some years ago Pi talked the European circuit owners into letting it install a monster beacon at the start line of most significant race tracks. Pi and other some other systems are triggered by this and, if this applies to you, you can skip the next few paragraphs. Owners of GPS-based systems are also spared the beacon problem.

The rest of us have the chore of taking our own timing beacon to the track, placing it correctly and then remembering to take it home again. Before Pi, the best place was the start/finish line because then the logging system and the official time keepers used the same datum point, and any discrepancies would be shown up at the end of the session. It also gave a fixed and unalterable reference point that everyone in the team understood. You could be sure that the beacon was in the same place every time that you ran at that particular circuit. This advice still

10.1. A photo like this identifies the chosen spot to place the timing beacon. Care was taken to include the pit number in the shot.

holds good for tracks where there is no Pi beacon.

Sadly, the Pi beacon is so powerful that it can 'drown' the output from any other individual beacon, so it becomes impossible to use your own beacon within several metres of the start line, and you'll have to find somewhere else. Choose somewhere near the start of the pit wall, and accept a few hundredths differences from the official times. When you've chosen the spot, take a digital photo, laminate it, and put it amongst the race-day paperwork to make sure that everyone knows where the beacon goes.

Some manufacturers provide coded beacons so your system can only be triggered by your beacon. Others rely on setting a minimum gap between triggers of 10 to 20 seconds, so that the system is triggered only once per lap. This still means that there's a good chance that your system will be triggered by the first compatible beacon on the wall, not your own. This is why the advice is to place your beacon at the start of the pit wall.

If things go wrong, you can usually

use the software to put a virtual beacon at the usual spot, but it's one more problem that can be avoided. Since you'll normally be competing against the same group of people at each event, you could come to some arrangement about sharing the beacon chores. In the long run, though, if you want absolutely no foul ups, it's better to take responsibility for the beacon yourself. You need to be sufficiently organised so that the other teams are happy to rely on your beacon. That way they won't start putting their own beacons where you don't want them. This is enough messing about to make having a GPS-based system look very attractive.

The beacon receiver

Monster beacon or not, the car's on-board receiver needs to be on the correct side of the car to see the beacon. In the UK we need to swap the receiver to the left of the car for Donington Park, Castle Combe and Croft because the pit lane is on the outside of the circuit. For all others, it's on the right side. So, not only does the switch need to be carried out for these circuits, it needs to be reversed for the next outing. One more item for the check list.

It sounds blindingly obvious, but the logger needs switching on every time it goes on track, so we need a procedure to make sure that it happens. Let the driver take responsibility. The whole team should know how the logger is actually started because, for some systems, merely powering up starts the logger, but for others for others the logger has to be deliberately started.

The health check

The immediate benefit comes from being able to reassure ourselves that the car is in good health. Chapter 4 suggests that one of the standard screens that you should prepare is one that shows the engine rpm, temperatures and pressures for the entire session and also a channel report that shows the maximum rpm used. In this way you can look at the entire run and be reassured that everything was in order.

Chart 10.2 plots engine characteristics against time. It shows the warm up period (on the left) and the way that the temperatures climb during the period of slow running and the pit stop. Cooling is adequate but marginal. Oil pressure starts quite high but stabilizes as the session goes on.

The logger can also tell you about how effective any radiator masking is. Run the car with and without masking and look at water, oil (and possibly also air inlet) temperatures, and you can see the impact that the masking has had. It doesn't matter too much if you can't run these tests back-to-back because what you're interested in is the temperature drop as a result of the masking. So long as you note the ambient temperature and the coolant temperature you can work out the temperature drop and use this in your decision making in future.

To estimate the temperature drop, look at the difference between the ambient and engine temperatures before and after masking.

The table shows that the practice run was done at an ambient temperature

	Ambient	Water	
Before	29°	67°	38°
After	34°	76°	42°
	Difference due to masking		4°

of 29°, and this gave a water temperature of 67° By the afternoon, the ambient temp had gone up to 34°, so it would be reasonable to expect the water temperature to be 72°. Instead, it was 76°, so 50mm (2in) of tape made a 4° difference to the water temperature. Further work showed that this was pretty well linear in the range that we needed to use. The figures are in degrees Celsius, but the logic holds good for Fahrenheit.

One of the great imponderables before the advent of data logging was the worry about whether an engine had lost its edge. The only measures that you could use were surrogates, such as leak down and compression tests. Now, with logged speed, it's possible to get an idea of the relative performance by examining speed traces. In theory it ought to be possible to log the rate of change in engine rpm and see any deterioration from this, but this does not turn out to be very satisfactory.

It's better to examine speed traces from different sessions (or even different

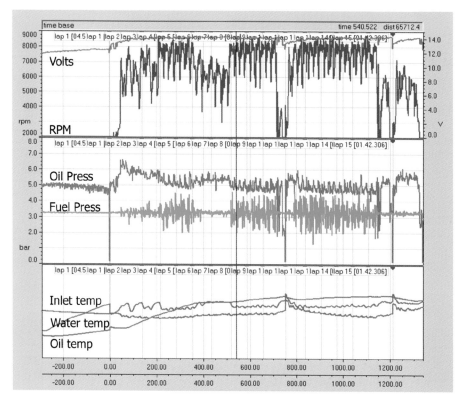

10.2. A single screen showing the engine health.

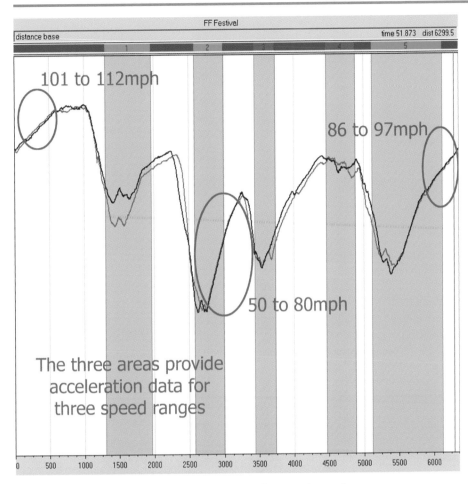

101 to 112mph

86 to 97mph

50 to 80mph

The three areas provide acceleration data for three speed ranges

10.3. Examining speed traces to evaluate engine performance.

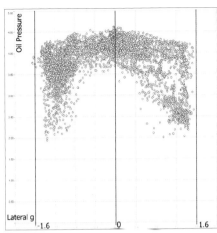

10.4. Oil pressure plotted against lateral g. As the g forces increase, the pressure falls away. Time for some work.

days if the conditions are similar) and see whether the slopes are comparable. If there really is a power loss, simply looking at the slope of the traces will confirm your suspicions. You can resort to more detailed measurement by selecting parts of the data and using pencil and paper and measure the time taken to get from one speed to another. This is pretty labour intensive, and the vagaries of track conditions make it difficult to get really accurate results. It would be usual to see differences of two or three tenths in the times that you measure. In the speed ranges shown in Chart 10.03, the dark blue trace was from early in the day, and the purple one was at the end. Black was always one or two tenths faster than purple, but all that could be said was that the conditions might have been slower later in the day. There was certainly not enough evidence to support the view that the engine was down on power.

DIAGNOSIS

If there are problems with the engine, diagnosis is a matter of common sense. If a fuel-injected engine misfires, one immediate test is to look at the fuel pressure trace and see that it stays within limits. If lack of power is the problem and you're logging ECU values, there is a whole range of diagnostic information available. Typically, you should have access to the battery voltage, lambda (air fuel ratio), manifold pressure and temperatures for inlet air, fuel and possibly exhaust. Plot these with rpm and throttle position and see what they tell you.

When first running a car, it's a good idea to plot the oil and fuel pressures against lateral g and long g to get some feel for the extent to which there is either oil or fuel surge. Chart 10.4 shows oil pressure vs lateral g, and is probably at the limits of what would be acceptable. There is a definite fall off in oil pressure at high g forces which, although it stays above 2 bar (28psi), is a cause for concern. A properly designed dry sump system does not show any loss of pressure in the turns.

RECORD KEEPING

The logger takes care of many of the record keeping chores (it provides lap and split times, and can tell you how many hours you put on the engine or miles on the chassis) but it imposes extra chores as well. The business with the beacon is just the start. Having a logger makes it even more important to keep records of the settings used at particular events. There's nothing worse than being unable to achieve previous performance

10.5. A cheap way to measure wind speed.

levels because you don't know precisely how you did it.

You'll also need to record ambient temperature and wind speed. Both can affect how the car works, and if you don't keep a record then you won't understand why the data says what it does. Ambient temperature is not such a problem because many loggers or ECUs measure ambient temperature automatically. If not, you ought to pack a thermometer. Wind speed is more important, it can explain differences of several miles per hour in terminal speed at the end of the straight. Wind speed is easily measured using an electronic anemometer. These are available very cheaply because they're a must-have accessory for anyone flying a model plane. Try your local model shop or do some internet shopping. While we're at it, there's a case for measuring atmospheric pressure as well, to put engine performance into context. In fact, throw in a cheap portable weather station so that you can record ambient pressure and humidity as well. This might help explain engine performance.

To use the logger to record running time, create a maths channel that will record either the number of miles run, the number of hours usage, or add in some sort of condition, such as the number of hours the engine spent above

a certain threshold, such as 3500rpm.

The function that calculates this is the 'integral function' and a simple formula to calculate mileage would be:

Mileage = INTEGRAL (speed)

You might need to adjust this to get the units that you want because the logger will measure speed in a particular way, for example, as feet per second, or as metres per second. This means that the logger might give you an answer in feet or metres. If so, it's a simple exercise to divide the result by 6360 (feet to miles) or 1000 (metres to kilometers).

To measure running time the formula will look something like:

Running time = INTEGRAL (if (rpm > 3500, 1, 0)) ÷ 60

It needs a bit of explanation. The 'if' statement checks to see that the rpm is greater than the chosen threshold. If it is, it returns the value 1, and if not records the value 0. Using this if statement on its own would provide a data trace that steps up every time the engine was used in anger, and step down again every time the rpm fell. This would be pretty to look at, but not much use. Using the INTEGRAL function calculates the area under this curve and, more usefully, shows this as seconds. The division by 60 gives us the answer in minutes. Using 3500 as the test means that we're only measuring the time that the engine is being run in anger. If we substituted a number slightly lower than the idling rpm (say 400) we would be recording the time that the engine was actually running.

This technique can be used for

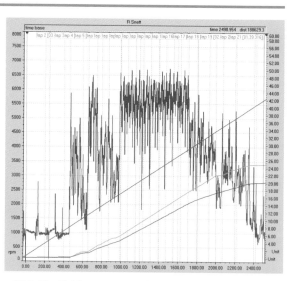

10.6. The light green records the engine running time, the dark green the time spent over 3500rpm. The red line records the distance covered.

'lifing' any component by deciding on the important characteristic. Professional teams use it for lifing the individual gearbox ratios by looking at the amount of time that individual gears are engaged. The expression below tests to see if the gear is 2 and adds up the time and so can be used for lifing second gear.

Second gear usage = INTEGRAL (if (gear = 2, 1, 0)) ÷ 60

The end of session data for engine or any other 'lifed' component is best collected using a channel report that shows the maxima, rather than drawing charts, but the end result is the same.

Chart 10.6 shows the rpm and running time and mileage for a race on one screen, and includes the warm-up, some time behind a safety car, and the roll back. The chart is only shown for interest; the significant figures are the final values. The engine ran for about 40 minutes in total, of which 24 were above 3500rpm, and 20 miles were covered. This can provide a basis for building up a record of how much work the engine has done.

TEAM MANAGEMENT

Chapter 6 gives strategies for getting the most out of the data in the shortest time, but managing the team, the car and the driver is vital to a successful day's sport. Emphasis has been on splitting the tasks between the notional roles of data engineer, mechanic, team manager, race engineer and driver, even though it would be a large amateur operation that had all of these people.

Just as someone has to step up and take responsibility for the logging system, the engineering of the car also needs supervising. What needs to be done and what the priorities are have to be decided. Having access to data doesn't replace your existing engineering skills, and what worked for you and your team before the logger will still work. The data gives us insight into what is going on, but there's still a need to translate this into plans of action.

The idea of splitting the corners into phases still applies, and it's still worth thinking about what inconsistencies show up, and in what phase. This is the key to starting to find explanations. Different people split the corner into different phases, but three or four seem to be a workable compromise:

End of braking area
Turning in
Steady state corner up to and beyond the apex
Exit phase, including the application of power

It's possible to argue with this set of stages, but all we're really interested in is the impact that each phase is likely to have on the behaviour of the car. The steady state phase is often anything but steady. In a slow corner, the driver is trying to get the power down as early as possible, but might not be able to manage it until quite late on. In a fast corner the driver should have enough confidence to get back on to the throttle as soon as the car is turned in. The remedies are the same, with or without a logger.

Thinking about what's happening to the car on track points you towards which data to look at for solutions. At the end of the braking area we can look for the way that the driver gets off the brakes and starts the transition into the corner. Brake pressure and balance data are useful here, as is the combined g. The turn-in phase can be illuminated by these charts, and the ones mentioned in the next sentence, to explore the steady state phase. In the steady state phase of the corner, a driver input screen that shows rpm and speed, steering and throttle and lateral g would be useful to assess the impact that the driver is having.

This is a time when egos can intrude. The first time that you look at the data there's a real tendency to expect it to confirm the preconceptions that you had about the car and the driver, and this leads people to look for facts to support their theories. Much better would be to look at the facts, and then develop a theory as to what is happening, rather than to have a theory that you look for facts for support. That way you can miss things.

One final point is that there's always a big temptation not to look at the data after the last run of the day. Everyone just wants to pack up and start the journey home. This is a pity, because it's always best to look at things while everyone is still together and while events are still fresh in peoples' minds. It might be a good idea for whoever looks after the data to be excused packing up duties in favour of making the coffee and presenting the important points of the data from the final run.

Whatever happens, you must not forget that competitions are driven by a timetable, and missing any of your time slots will have consequences that no amount of data will get you out of. So it's important to realise that the logger fits in around the event, not the other way round. Just like the engineering, the logger does not alter the way that you manage the day; it just makes it more effective. It creates another set of tasks to perform, but this is not all counter-productive. The immediate download of data after a run lets you set priorities by showing up problems and highlighting opportunities to optimise things like gear ratios. It's a useful peg upon which to hang the debrief, and involve all team members in the running of the car.

TEST DAYS

Test days are similar to, but different from race days. A plan is vital, otherwise there's an inevitable tendency for things to drift and for this to be justified as the driver getting more 'seat time'. Even if there is a cast-iron case for the driver clocking up the miles, simply flogging round without experimenting is a wasted opportunity. Even if you only change the settings of things like roll bars or ride heights, it can give the team an insight into the effects of these adjustments for some future competition. A worthwhile experiment is to wind all the adjustment off a single damper, so that the driver has some idea of what a malfunctioning damper feels like, and the data engineer can see how the data is affected. If 'seat-time' is the goal, at least make the effort to learn more about the car and how it needs to be driven.

Much better than this would be a program of changes with some specific objective in mind. Given that there is a program, the logging can be directed towards understanding the results of the program. So, if the objective is to make the car less nervous in the corners, the data engineer can pre-prepare screens that might show this up. The multiple overlay technique will

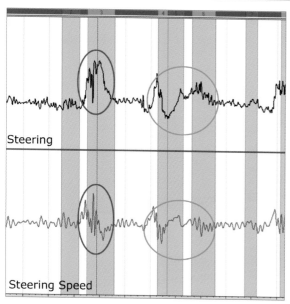

10.7. A spot of research before a test to identify steering speed traces for instances of good handling (green), and difficult handling (blue).

still work, but additionally it might be worthwhile looking at the derivative of the steering channel (steering speed) to see how quickly the driver is having to work to stay in charge of the car. If you know that this is going to be part of the test programme, then a bit of work with existing data will help you build a meaningful steering speed channel and show you what steering speed to expect in a corner where the car is well balanced, and where the car is known to be uncooperative. It will also help you work out whether the data needs filtering to make it intelligible and, if so, what technique to use.

How the logging works depends on the format of the test day. If the track has been booked for the whole day and there is a full programme planned, the data will need to be downloaded after each run and analysed while the next run is taking place. This means that there is a real need for accurate record keeping and things will work better if a programme is devised before leaving home. If the test is in sessions with a run and then a wait for the next session, this makes things easier. In all cases it is important to try to stay on top of what the data has to tell you because it might be that it gives insights into new directions to take and will cause you to want to throw away the script.

It's the same for test days as it is for race days, the techniques that worked before data logging will still apply. If at all possible, you should try a setting, go back to the original, and then try the test setting again in order to reassure yourself that the test is valid. If you believe in the level or layers approach of:

Basic settings
Dynamic behaviour
Aero
Gearing

and this has served you well in the past, there's no need to abandon it just because there is logged data available. In fact, what worked for you before the logger will still work for you now.

AFTER AN OUTING

The data should also be looked at again without the inevitable time pressures found at the circuit. It might be that the team is so well organised and the car so well prepared that there is nothing much to do on the day of the event or the test and, if this is so, it's important to use the available time for a more leisurely look through the data. The first attempt got us a general impression of how things were going, and helped fill out the job list, but a second, third and fourth attempts can be more considered, and either structured or unstructured. The unstructured approach would just browse through the data to see what you see, where the structured approach would have a specific goal in mind – such as:

Is there a difference between high speed and low speed corners?
Would gearing changes help acceleration?
Is there evidence of brake fade?
Is the start procedure optimal?

An unstructured approach would be looking for interesting facts or relationships. It's a time to experiment with using the data in unconventional ways. Generally, though, this sort of relaxed look at the data will take place away from the track. It's a Monday morning job for the professional team, one evening in the week after the event for the amateurs.

Chapter 11
The driver

If Steve McQueen was right, and racing is living and everything else is just waiting, it makes sense to spend some of the waiting time with the data. For a driver, the whole point of logging is to go faster, and, as this can be at least a three phase operation, this chapter is split into 'before the event', 'during the event', and 'after' sections. The emphasis tends to be on what the data can tell you on the day, and this is obviously a key function, but don't forget that even before the event, the data from the last time out in the car, and the previous outing at that track, can be valuable reminders of what goes on. You must also take time for a really thorough look at the data after the event, but while things are still fresh in your mind. In this way, you stand a chance of uncovering the secrets locked in the data, but which are temporarily obscured by the pressures of competition.

BEFORE THE EVENT

Review the data from this track a few days before the event, to remind you of what happened last time out. The software will provide a list of lap times and detailed information about speeds, rpm, gear ratios, shift points, and a hundred and one other things. You don't actually need logged data to do this, because, as an intelligent driver you'll have brought all the important points together in notes written during the last event and at the debrief afterwards. But, having the data adds another dimension to your preparation; since the last outing at that circuit it's likely that things have changed, and part of the preparation should be to evaluate the effect of those changes on the current event.

One thing that the logger will do for you is to print out a track map. You can then annotate it with information about

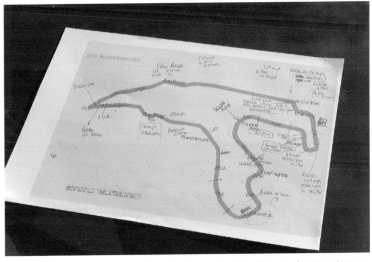

11.01. The logger can provide a track map for the driver to keep notes on.

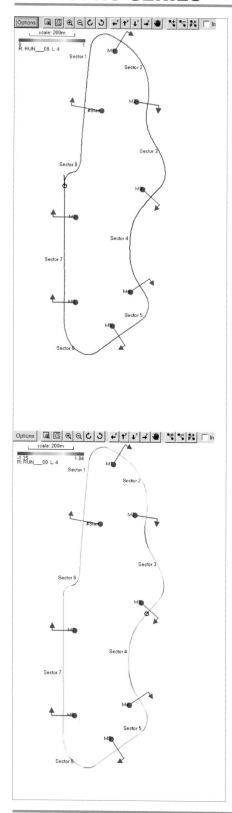

11.02. Two track maps showing different information about the track.
The top one shows positive and negative long g – whether the car is braking or accelerating. The blue zones indicate the braking areas, and also give an indicator of gearchange points.
The bottom map is shaded to show lateral acceleration, and indicates how hard the corners were being taken.

the lines, the braking points, gear ratio, and expected revs. In fact, it's well worth printing off several copies of the track map to take with you. They'll be useful for updating the existing map, for the debriefs, and for communication and explanation purposes.

Other information can also be recorded that will not only help remind you, but can improve your understanding of the data. Bumps, poor surfaces, track cambers and so on all show up in the data with an impact on many of the channels, and knowing about these things can help understand why a particular piece of data is not as expected.

A typical case here is the extent to which the driver uses the kerbs. If you're in the habit of getting on to the kerb on the outside of a right-hand bend immediately before turning in (to maximize the corner radius and optimise the line) this can show as a brief instance of lateral g on the accelerometer. Anyone reading the lateral g trace would see this as the car moving to the left and might assume that the driver is flicking the car left to get it to turn right more convincingly, and might worry about lack of bite in turn-in. But the clue to the correct interpretation is that all this happens with very little steering input, so that the flick explanation could not have been correct. Easier to know this from a track map with notes on than to waste time finding and verifying an explanation the hard way.

Similarly, using the kerb at the apex of a corner will show as a small increase in the lateral g trace. It's fairly easily recognisable, being of very short duration, but again, it's easier for the driver to provide the explanation than to have to work out the amount of kerb that the driver is taking from the size of the bump.

The notes are a starting point, but the analysis software lets you look at the data in detail. This will enable you to visualise the shape of the track and the speeds that you should be achieving. A good way to get a quick overview of what happens at a particular track is to use shaded track maps. If the software allows it, have a look at a range of maps that show the track shaded according to such things as speed, mph/1000rpm or gear, and throttle position. In this way, you can remind yourself about the speeds that you could achieve, the gear ratios you used, and which corners were taken with wide open throttle and those where you needed to be a bit more circumspect. The throttle position map also gives some clue as to braking points. The two coloured maps shown in Chart 11.02 show the braking flag (whether long g was above or below zero – red for accelerating and blue for losing speed) and lateral g to separate the on-the-limit corners from the rest. A few minutes playing with this feature will refresh the memory on many aspects of driving the track.

Once the memory is refreshed about the general business of driving the track, time to go into a bit of detail. A good place to start is braking points. Plotting long g trace and corner radius, or lateral g against distance, will help you identify whether corners are early or late apex and precisely how many metres before the apex you start to brake. This latter figure is not much use, but it does give you a starting point to relate it to the countdown boards on the track. Similarly, target rpm, gearchange points, and cornering lines and speeds can all be reviewed. This is where video is invaluable; it will tell you more about what a circuit looks like in a way that no amount of walking the course can do. It also gives a pretty instant view of cause and effect with driver inputs. The objective is to go to the track knowing what to expect, and having some sort of idea where improvements can be made.

AT THE RACE TRACK

Properly prepared, you should be able to minimise the time spent familiarising yourself with the track, and get on with the business of driving quickly. After the run you need to review the data and compare it with previous events. Chapter 6 tells you how to review the data as efficiently as possible. Patience is needed here, because the first priority in reviewing the data is to identify what needs to be done to the car before the next run. When your turn comes, the first thing to look at is the list of lap times. The fastest time is always interesting, but the spread of times is probably more significant. There is something very satisfying about the 'banzai' qualifying lap, and it's worth looking at it in detail, even though, by its very nature, it's not something that you would want to try to repeat on every lap throughout the race.

Generally, a consistent list of lap times is what you're looking for, with the times early in the session perhaps being a couple of seconds slower than the absolute fastest lap, and then working up to a time within a second or so of the fastest. Obviously this will vary according to circumstances, and the first time at a particular track will show much greater variations than when you return to a track that you know well. The obvious first candidate for scrutiny is the fastest lap, but fascinating as that lap is, don't forget the others. Unless the lap was perfect, there'll be parts of other laps that were quicker in certain sectors, and this need to be looked at too.

Split times

Start with the lap and sectors report that shows lap times and sector times. Generally, these will also calculate the fastest theoretical lap and, while this is great in theory, don't forget that there's a good reason why some teams call this the fantasy lap. By all means enjoy the thought of the fastest theoretical lap, but pay more attention to the fastest rolling lap.

Traditionally, laps were split into much longer sectors – three or four to the lap – and this can have its advantages. Longer can be chosen to reflect different aspects of the track. For example, somewhere like Spa in Belgium splits conveniently into a fast first sector on the old circuit, a technical middle sector with the majority of the bends, and then the high speed return to the start/finish area. The first and the last sector require low drag, low downforce settings, and the middle part needs higher downforce to cope with the corners. Splitting the lap into three sectors allows the team to understand and balance the conflicting demands of the various parts of the circuit.

As driver though, you need to understand your strengths and weaknesses over individual laps and over a session as a whole. If, over time, the fastest rolling laps all tend to start

at the official start/finish line, then this is evidence that you're the type who tends to screw up the level of effort and go for a 'split-or-bust' frightener of a lap. The other way of looking at this is that there is a certain amount of cruising going on that is possibly a waste of track time. Would it be better to try to be always on the case and looking for the chance to improve? Generally, fast times happen toward the end of a session, but if they're clustered in the early stages or towards the middle, there might be a lesson. There are lots of possible explanations, such as lack of traffic, track condition, or fatigue, each of which should tell you something about how to approach the next session.

If the fastest times are clustered together, then you know where to direct your attention. If they're spread across the table, then you need to look for fastest sectors and other laps where similar sector times were recorded. This should point you in the direction of examples of driving a part of the track well, and other data should then be explored to see what you do that works. The point is not just to concentrate on the three or four fastest laps, but to dig around and find lessons in the other laps as well. Chart 11.03 used the split times to find the best times through two sectors, but the interesting thing is that

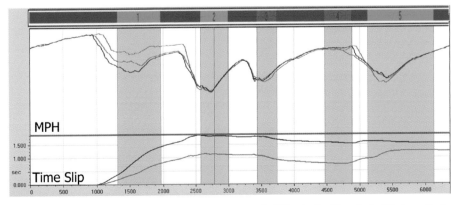

11.03. The red and the blue lap were over a second off the fastest lap, but still show that time can be saved in Turn 3.

If there is ...	Then ...	
Understeer on turn-in and ...	Mechanical grip	Aero grip
... it won't point	blame the driver	more splitter
	less front bar	
	softer front springs	
	lower front ride height	
	increase front bump damping	
	increase camber	
... it points but fades	blame the driver	more splitter
	increase toe in	
	increase front bump damping	
... it is unstable at turn-in	blame the driver	
	less front bump stop	
Understeer mid-corner and on exit	blame the driver	more splitter
	reduce relative front roll stiffness	lower spoiler
	reduce front droop damping	
Oversteer on turn-in	blame the driver	
	check for rear toe-out	
	look for something broken	
Oversteer mid corner and on exit	blame the driver	increase rake (as an aero fix)
	reduce rear roll stiffness	raise spoiler
	reduce rear camber	

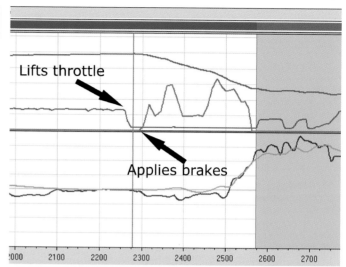

11.04. Evidence of a problem (the driver lifts off the throttle but waits 0.3sec before braking), but is the car or the driver at fault?

they came from laps that were around a second off the pace. Generally, laps like this would be ignored. They show clearly that there is a little bit of time to be saved on Turn 3 that pays dividends all along the straight to Turn 4.

Driving style

When a good performance depends partly on the driver and partly on the car, it's only natural for each party to blame the other. Natural, but not productive unless it's done in the right spirit. Many race engineers keep an 'if-then' list to prompt them in the heat of competition.

This one always starts with 'blame the driver'. This is not simply a joke or a tactic to shift the blame away from poor engineering, but a recognition that almost any symptom can be caused by driving technique. Sometimes it's just poor driving; arriving at any corner much too quickly will always promote understeer. What seems like poor driving technique, however, can be brought about by trying to drive around problems caused by poor engineering. The reason the driver arrives at a turn at impossibly high speeds may be because the car is so slow everywhere else that he or she is trying to compensate in any way possible. Hence the turn-in understeer that might be completely different when driven under less pressure.

So, when reviewing the data, it's vital to be honest with yourself and with the others in the team. There's often a testosterone-driven tendency for people to reach an early diagnosis from the data, and then spend time arguing for the original diagnosis, not testing it further to see if it's a good one. People come up with a pet theory, and then defend vigorously until absolutely proved wrong. This tendency is particularly

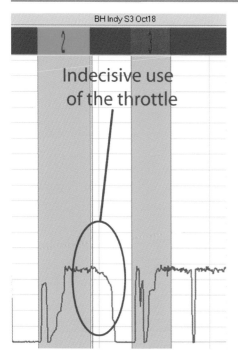

11.05. The driver is indecisive about when to come off the throttle here.

likely in the discussions between engineer and driver – 'It's not my driving that's at fault, it's the car you provided me with ...' – so you really need to be aware of it and make sure that egos don't get in the way of honest analysis.

Chart 11.04 shows a typical problem. Turn 3 at Castle Combe in the UK is a fierce combination of high speed uphill approach, a fast left turn over a brow, and g-forces throwing the car to the wrong side of the track for a long right-hand hairpin bend. The data shows a measurable gap (0.3sec) between getting off the gas and onto the brakes, and this was visible in every lap of the session. Is this the driver's self-preservation instinct kicking in, or is the car just too lively in this complex for him to have complete confidence? Chart 11.05 shows the same symptoms at a different circuit.

Cornering

The datalogger can provide insights

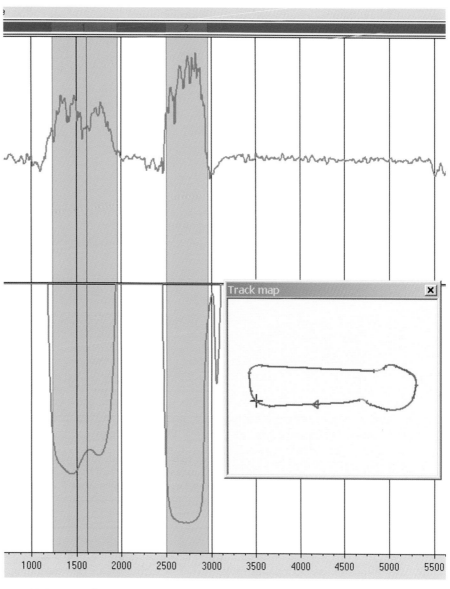

11.06. Lateral g (pink) and corner radius (green) both indicate problems with the racing line. In the first turn (first shaded area) the driver released the steering and reduced the corner radius. This bend is faster when treated as a single constant radius curve. There is another (unexplained) adjustment to the line at the very end of the second turn.

into cornering style and cornering effectiveness. Chapter 7 on handling and steering deals with steering input and corner radius in some detail, but much can be learned from a plot of the corner radius. It is tempting to see the corner radius curve as a map of the line through the corner, but it isn't.

Figure 11.06 shows lateral g (pink) and corner radius (green). Turn 1 is taken as a double apex bend – the tighter radius of the first part is eased in the middle of the corner, and then tightened again towards the end. Turn 2 shows a rapid turn-in (indicated by the steep initial phase of the turn), a period

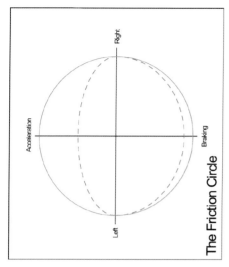

11.07. The friction circle.

of almost constant radius in the middle phase, followed by rapid release of the steering and a strange little adjustment at the end of the turn. This trace can be useful for recalling what lines were actually taken through a corner.

Using the tyres

It is the driver's job to maximise the grip that the tyres provide, and the usual way of demonstrating this is by use of the friction circle. In case you're not familiar with the concept, the theory goes that, for any car, it's possible to plot a circle which shows the absolute limits of grip, and a good driver will be able to operate at the limits. The size of the circle varies from car-to-car (Formula 1 might be looking at 4 or 5g, a roadgoing saloon would be good if it reached 1g) and it's unlikely to be truly circular.

Diagram 11.07 shows a typical, theoretical friction circle. The green circle represents the maximum grip that a tyre can give. If we assume an arbitrary value of 1.5g, the idea is that the car can achieve that grip in all directions. But not only can it give the grip under acceleration, braking and cornering, it can do it in the transition phases. This provides us with a justification for staying

on the brakes after the car has been turned in, and also reflects the reality that any real racing driver will get on the throttle long before the car is out of the corner.

The red line is closer to reality. The bottom part of the red curve represents braking, and is about 85 per cent of the 1.5g. This is usual, and is a function of the shape of the tyre contact patch. A road car with a square contact patch generally seems to have about equal braking and cornering power. A racing car with a contact patch that is wider than it is long will have less braking grip than cornering power. There isn't a fixed definable relationship, but the rule holds good in most cases, and it helps in interpreting friction circles created using logged data.

The top part of the red curve represents acceleration, and, in this case, is about half of the 1.5g of the full circle. This is a power and weight problem, and has nothing to do with the available grip. With more power and the torque multiplication that comes with low gearing, this tyre should reach the 85 per cent that was possible under braking.

On the X axis, the left and right limits will be represented by the steady state cornering potential. On the Y axis, the positive values will be limited by the maximum rate of acceleration, and the negative values by the maximum braking ability. So, the starting point is that the driver will be able to reach these maximum values. But, better than

that, he or she will be making most use of the tyres in the transition phases from braking to cornering, and from cornering to acceleration.

Going through a corner, the first thing that shows on the friction circle is that the g force plunges downwards as maximum braking is achieved. If the driver then gets off the brakes and turns in, the trace will climb back to zero, and then head off sideways as cornering force is built up. At the exit of the corner, the lateral g trace moves back to zero and the throttle can be opened and the long g trace will head upwards. With this sort of cornering style, the friction circle is more like a friction cross and indicates that there is much more grip to be extracted from the tyres by exploring the perimeter of the circle.

Turning in on the brakes will move the trajectory around the outside of the circle. What is happening is that there is a transition from longitudinal to lateral g. The driver is not using absolutely all of the available grip when he or she is

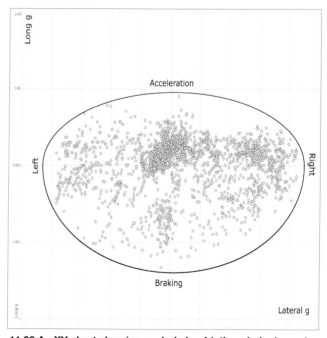

11.08 An XY chart showing a whole lap friction circle drawn by the logger.

coming off the brake pedal. So, some of the grip could be used for turning into the corner. This is quite convenient really, because the turn-in phase of the corner is transitional, gradually building up cornering force as the tyres develop their optimal slip angle. So, if the driver can judge the braking so that he or she still needs to shed a small amount of speed while the car is being rotated into the corner, the tyres can provide the grip.

The logger will be able to provide an XY chart with lateral g on the X axis and longitudinal g on the Y axis. This provides you with a real friction circle as opposed to one from the vehicle

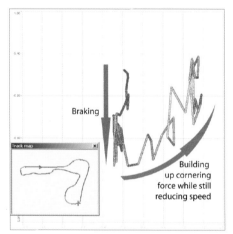

11.09. Looking in detail at a single corner friction circle.

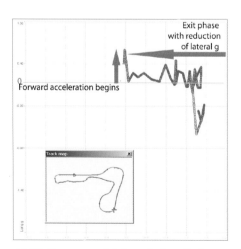

11.10. One corner friction circle exit phase.

dynamics text books. You can present this in two forms; one for a complete lap, or even the whole run that shows the general use that is being made of the available grip, and another for a single corner or sequence.

This sort of friction circle provides a general commentary on the driver's style, but it also reflects the nature of the car. A small single-seater should respond well to turning in under the brakes and accelerating long before the exit of the corner. A powerful, heavy car might only be capable of being driven in a series of point and squirt actions, and driver survival might depend on slowing down in a straight line, negotiating the corner, and waiting until the car is settled before getting back on the gas. Chart 11.08 is for a whole lap, and shows that improvement is possible. Although the turn-in phase seems to blend braking with cornering, there's very little evidence of acceleration out of right-hand turns.

Using a single corner trace is also useful. If the software allows it, select the corner that you're interested in and create an XY diagram with the data points joined by a line. This will show your progress through the corner in terms of getting the best from the tyres, especially during the transition phase where there can be surplus tyre grip available that you're not using. You can see the initial braking effort, the transition into the corner, whether any adjustments had to be made, and the commitment to the exit phase. Charts 11.09 and 11.10 show a typical set of traces. The blending of the brakes and steering on turn-in are handled well, but there's no

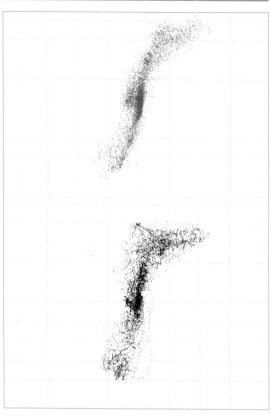

11.11. Lateral g (Y axis) vs steering. The two traces are indicative of the amount of effort shown by the driver at two different tracks. Everything seems to be slightly more under control with the red data than with the black.

evidence of forward acceleration until the car is running in a straight line.

How hard were you trying?

There are days when everything is just easy, and fast times come with very little effort. On the other hand, there are days when no matter how you try, competitive times just don't seem to be there to be had. The data can show when things are difficult. One good indicator is to plot steering on the X (horizontal) axis and lateral g on the Y axis. Some people suggest that this is a measure of the tyre's slip angle, and if you squint your eyes a bit you can see the relationship between the two curves. Interesting, and probably useful if you have definitive data about the real slip angle, because then you could evaluate the amount of time spent operating outside the proper operating envelope. In the real world that club competitors inhabit, it is a bit difficult to see what use can be made of it. On the other hand, creating this plot for five or six of the fastest laps in a session can give some insight into just how hard the driver is trying.

The two XY graphs in Chart 11.11 are both taken from the same car, with only minor changes in set-up, and in a period when the car wasn't going particularly well. The red figures relate to a track which is only visited once a year and is not a favourite of the driver. On the other hand, the black figures were taken at a track that he regards as his 'home' track. The difference in shape is subtle, but real. The black trace shows lots of occasions when the car was outside the expected performance envelope. The explanation we reached was that at the red track, expectations were low, and the performance shortcomings accepted as inevitable. At the home track (the black figures) the driver expected to be at the front and was driving much harder to compensate for the inadequacies of the car. This difference in expectation and effort was reflected in the significantly different shapes of the two sets of data.

Coasting

When thinking about how hard you were trying, one thing to consider is coasting time. This unforgiving trace lays bare the amount of time spent each lap with the throttle shut, and not yet on the brakes. One way to show this is display a speed trace every time that the car is coasting. If the speed trace is measured in metres per second, the 'integral' function can be used to calculate either the number of metres per lap, or the number of seconds per lap that meet the coasting criterion.

The maths is fairly complex because you'll be relying on an 'if'

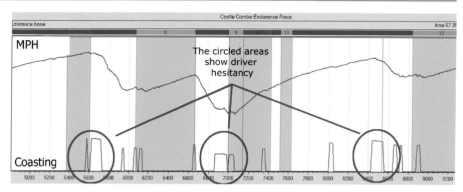

11.12. The coasting trace (red) records when the driver is off the throttle and not on the brakes.

statement inside another 'if' statement. 'If the throttle position is less than 50 per cent, and if the brake pressure is less than 2bar (30psi), then display the speed, or else show zero. Depending on software, that would translate into a formula something like the one below.

Coasting = if (throttle < 50, if (brakes <2, speed, 0),0)

The syntax for most if statements does not actually include the words then and else, but just assumes that the then will start after the first comma and the else after the second. Reading the above formula from left to right gives us if throttle less than 50 per cent, then if brakes are less than 2bar, then record the speed, or else return zero. The previous part of the expression is the then part of the first if, and, in fact, because it is itself an if, it has its own then and else inside its brackets. So, now that the first if and the first then are dealt with, we need to go on to the else that belongs to them and that is the final ,0) of the equation. If this makes your brain hurt, find an Excel expert, he or she should be able to guide you through it.

This gives a nice trace that shows the speed when the car was coasting. It graphically demonstrates the parts of the circuit where the driver is being indecisive. Chart 11.12 is an example, and picks out major areas of indecision

by the driver, each taking almost a whole second. Small upward blips are inevitable, either during gear changes or in corners when the driver gets out of the throttle, but there should be no marked 'flat top' areas.

If you found the logic statement an interesting challenge, then you can add another if into the nest and check to see that the car was not in a corner. So, if you're off the throttle, and if you're not on the brakes, and if the lateral g is between -0.8g and +0.8g, then put up the coasting flag. For most of us that may well be a refinement too far.

Using the controls

You can get a bit of a picture of your driving style by looking at the way in which you use the controls. For instance, it's possible to measure the speed at which you apply the throttle, or get onto the brakes, or turn the steering wheel, and these plots can be very revealing.

Throttle control

Virtually every text-book on how to drive a racing car stresses the importance of driving smoothly and not upsetting the car, or more importantly, the tyre contact patches. Typically, you're told to feed the power in gently, and you sometimes wonder if the author has ever sat in a racing car. Nothing seems very gentle at the time, but with a logger we can measure the vigour with which the pedal is hit as a change in the throttle position per second. Some people suggest that a rate of somewhere between 150 and 200 per cent per second is the maximum, and that fast drivers stay below this throttle speed. They don't jump on the pedal, they feed it in as quickly as the car will allow. So a throttle speed trace might be worth looking at. The maths is simple enough:

Throttle speed = derivative(throttle)

If this gives a noisy trace, go back to Chapter 4 and decide whether to use some method of cutting out the unwanted readings. In any case, the values will need to have the maximum level of filtering applied to them to give an understandable trace. The trace shown in Chart 11.12 cuts out any speed below 20 per cent per second simply to smooth the signal when the driver is fully on or fully off the pedal, and has the maximum filter applied.

If you think that this might be one of the areas for improvement in the driving and the numbers give cause for concern, it's an easy enough thing to do a few laps making a conscious effort to be more circumspect with the throttle pedal. The data will tell you whether there was a pay off in terms of speed. It may work, or it may not, but it will have added to your knowledge. Don't forget, though, that there's still the chance to lean on the engineers for a car that will react better to the more urgent approach.

Chart 11.13 shows throttle speed in the lower section of the chart. It's plotted in red against the throttle data in blue with a green line representing the threshold of 200 per cent. It shows that the driver is at the upper end of what is recommended, and might consider a change of driving style.

Steering speed

What works for the throttle works for the steering as well, although, in this case, it makes a pretty good indicator of where the driver has to work hard. Steering speed can be calculated in the same way as the throttle speed, but it does not need a dimension. It does not matter whether you show it as change in logger volts per second or radians or degrees of steering wheel or road wheels, it's the shape that's interesting. It still needs quite heavy filtering to provide a trace that gives any insight.

Steering speed = derivative(steering)

In Chart 11.13, arrows are used to indicate the areas for concern. The turns where the driver is not under much stress have green arrows pointing to

steering speed is high at some particular phase of the corner (turn-in, steady state or exit) then this is one more clue to how the car is handling.

Some analysts extend this idea into looking at the braking application as well. They construct an index of braking aggression by measuring how quickly you build up braking effort, with the idea that quicker is better, and they measure the braking finesse by looking at how gently you come off the brakes. It uses exactly the same logic as for throttle speed and steering speed, but needs a lot of filtering to make a readable trace. It definitely seems like one for dark winter evenings when there's absolutely nothing on TV.

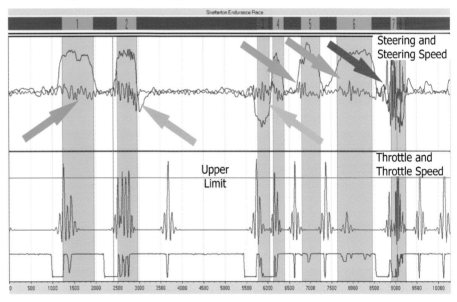

11.13. The raw data for steering (top) and throttle (bottom) are shown in blue, and the steering speed and throttle speed are shown next to them in red. See text for an explanation of the meaning of the arrows.

them. The orange arrows point to the turns where the driver is busier, and the red arrow points to the final turn of the lap where the driver is really earning his keep. If you relate the steering speed to the speed, you might get an insight into whether this is a factor in fast or slow corners. If there's evidence that the

Using the gears

The precise way in which you change gear will depend primarily on the type of box, and the data can help identify the correct technique because it's relatively straightforward to measure the time taken to change gear with a strip chart. You might have to nurse a production-based synchromesh box and get off the throttle, use relatively light pressure on the lever, and still have to wait for the synchro hubs to do their stuff. With a dog-clutch gearbox, simply putting a bit of pressure on the lever and relieving the load on the dogs by either hitting the clutch or momentarily lifting the throttle will make the change.

Whatever type of box it is, you can measure with some degree of certainty which is the fastest method. Chart 11.14 shows comparisons of two gearchanges, one with a lift of the throttle, and one without. Looking at the top part of the chart first, this is a traditional lift and shift approach. The way to measure this is to measure the time gap putting the cursor on Point A, reading off the time into the lap, and then doing the same for Point B. This amounts, in this case to 0.35sec. An alternative way would have been to measure the time that it took for the mph/1000rpm trace to make the step up from one gear to the next.

The lower part of the chart shows a flat shift where there is no characteristic vee in the throttle trace, so this method of measurement won't work. Instead, it's worth looking at the speed trace (within the blue circles). In the top half of the chart there's a small but distinct flattening out of the mph trace while the change is being made. In the lower part of the chart, this doesn't happen, so we assume that no time was lost during the shift. Don't take this to mean that you should always flat shift, but that you can learn what works best for your car by zooming in on a few gearchanges and evaluating which ones work best.

In the lower set of traces, the step up into the next gear (as shown by the mph/1000rpm line) is distinctly lumpy, indicating that the transmission in general is having a harder time. However, do you compete to preserve transmission components or to go fast?

As for when to change gear, that

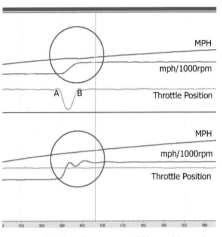

11.14. A tale of two gearshifts, with and without lifting the throttle.

depends on the gap between the ratios; there's no 'magic bullet'. Every combination of gear ratios for a car will have a different set of change up points. Chapter 8 deals with how to calculate optimal gearchange points, but it needs full rolling road data and a mathematical model. From the pragmatic viewpoint, the only way to find out is to try things. You should have a pretty good idea where the optimum point is just from the feel of the car. You can refine this by looking at the data.

If you have a long g sensor fitted, then some people suggest that you can identify the optimum acceleration in any gear by inspecting the curve – the long g will climb and then fall, and the peak

11.15. Checking to see whether higher rpm shifts make time. The time slip indicates that they do in this instance.

value represents the best time to change up. The problem with this method is that it doesn't take into account the gap between this and the next ratio. If there's a big gap, there can be a payoff in hanging on to the lower gear, just to put the engine higher up its power curve in the next gear.

A better method is to examine acceleration after the gearchange. Early in the session you should do a series of laps with different change points, say one lap at 8200, one at 8400 and one at 8600, and then get on with qualifying in the normal way. Looking at the data afterwards, you can see which was

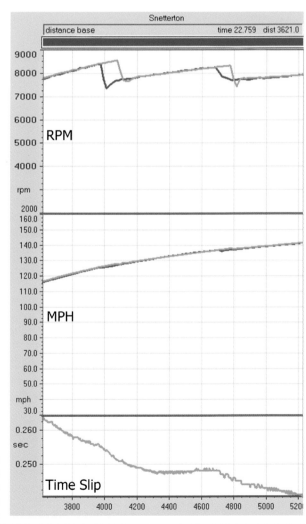

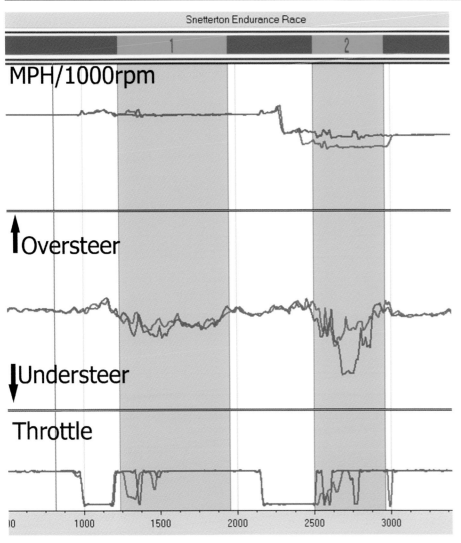

MPH/1000rpm

↑Oversteer

↓Understeer

Throttle

11.16. Using second gear (the red trace) reduces the understeer and allows earlier throttle application.

the most effective. In Chart 11.15, the orange traces relate to a lap where the driver used higher revs. The speeds are almost identical, but the time slip trace indicates that higher is better. The problem is that the amount of time saved is only a couple of hundredths, so what we take from this is that the shift point in this instance, with these ratios, is not too critical.

Which gear to use is not always a clear cut decision. Chart 11.16 shows a driver experimenting with different gears through a turn. The top trace shows the driver experimenting with a lower gear on the red lap. The trace below that shows understeer (when actual steering is greater than the required steering for the turn) and there was significantly less understeer on the red lap.

This is backed up by looking at the throttle trace which shows that the driver was able to get back on to the gas about a third of the way through the corner, rather than have a couple of attempts on the other lap. Using the lower gear was obviously the way to go, and this led on to discussions about whether it was worthwhile running a slightly higher ratio to avoid over-revving the motor.

AFTER THE EVENT

To get the best from your system, it's crucial to spend time with the data after a competition and while the events are still fresh in people's minds. A trip to the pub on the Wednesday after the race works best for us. Everyone has questions and pet theories, and it's a good time to brainstorm your way through the data. This often raises issues that lend themselves to a more considered approach on an individual basis afterwards. Whatever happens, don't just file it away until next time. There are secrets in there waiting to be unearthed.

Chapter 12
Other disciplines

This book was written from the circuit racing perspective, but most of the ideas apply to other branches of four-wheeled motorsport. Whether it's to do with buying, installing, or operating a system, the principles will be the same. What will inevitably change is the emphasis placed on certain aspects, because every branch of the sport has its own special challenges. This chapter covers some of the special circumstances in different types of competition.

HILL-CLIMBS SPRINTS & AUTOCROSS

These are lumped together since they have the common characteristic of short, intensive runs from a standing start, and they need everything to be optimised right from the green light. It makes sense to use the data to make sure that this is the case. Because of the importance of wheelspin during the start and under braking and acceleration, there is a much stronger case for logging all four wheel speeds than there is with circuit racing.

Calculating a combined g channel also helps to make sure that you're getting the absolute maximum from the tyres, and this will apply not just to turning-in under the brakes, but also the exit under power. If the tracks are bumpy, there's a case for logging vertical g to help understand why the car does what it does.

There are a few housekeeping issues with making a single run as opposed to repeated laps of the same track. If you use a beacon to trigger the logger, two are needed, one for each end of the run, so it's time to find a buddy with a

similar system. GPS is a help but has its own drawbacks. It will not necessarily be possible to overlay runs precisely when they're separated in time by a few hours. There is inevitably a small amount of drift over time and, while the effect of this is pretty minimal, if you're drawing in your own start and finish lines, a very small drift (say, 12in or 30cm) can

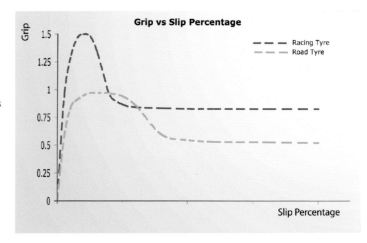

12.01. Grip is maximised when there's a small amount of slip at the contact patch.

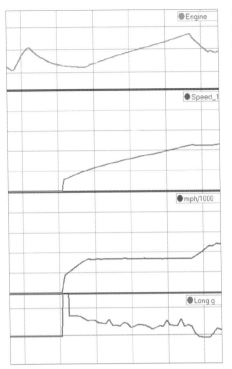

12.02. Because it takes time to trigger the sensor, the first part of a start is not properly recorded.

in diameter. This is resisted by the weight (so, sticking to imperial units), if the torque is 100lb/ft, the thrust will be about 1000lb at the tyres. If the car weighs 1600lb, the theoretical acceleration of about 0.625g (1000 ÷ 1600 = 0.625). If the tyres will cope with 1.5 g, that implies a coefficient of friction of 1.5 multiplied by the weight on the back wheels (say 800lb), so that they'll cope with a thrust of 1200lb. This car would be torque limited because the thrust is less than the grip, and the driver's job is to make sure that the maximum amount of torque gets to the back wheels throughout the launch phase. This will involve slipping the clutch to keep the engine rpm at the figure that develops peak torque until the wheel speed exceeds this.

If the car has more grunt than grip, and the thrust at the contact patch exceeds the potential of the two tyres, then the driver's job is to maintain the optimum wheel slip ratio. Generally speaking, racing tyres offer greater

coefficients of friction than road tyres, but do so over a smaller range of slip values. Chart 12.01 shows the shape of the curves, but the tyre manufacturers seem to want to keep the values secret. Logging the car speed, the speed of the driven wheels, and long g will allow you to find out what works best for your car, your tyres, and your track surface.

GPS-based loggers do a better job of recording what goes on at the start than those that use wheel speed sensors, because it takes a finite amount of time for the wheel to move and record a value on the logger. This results in a sudden step up of the speed, mph/1000rpm, and long g traces.

Charts 12.02 and 12.03 demonstrate the difference. In 12.02, the top green trace is the engine rpm, the purple trace is speed, and the red trace is miles per hour per 1000 rpm. The bottom green trace is long g calculated from the speed. This means that three of the traces on the chart owe their

compromise the accuracy of the data. Every competitor would like to launch from just behind the line because it uses up the time taken to wind up the transmission and tyres and get moving before the timing beam is triggered. You cannot do it in the real world, but you might be doing it in the virtual world of handdrawn start lines. Similarly, if the finish line is not quite in the same place, comparability is lost. These are not major issues, just minor grumbles. You'll not be able to rely on the data for absolutely comparable 0 to 64ft (19.5m) timings.

The exact technique for getting off the line will depend on whether the car is limited by traction or torque. The torque at the flywheel will be multiplied through the gear train to about ten times its original value by the time it reaches the tyre contact patch. These are broad figures based on a 2.5:1 first gear, a 4.0:1 final drive, and a back wheel 24in

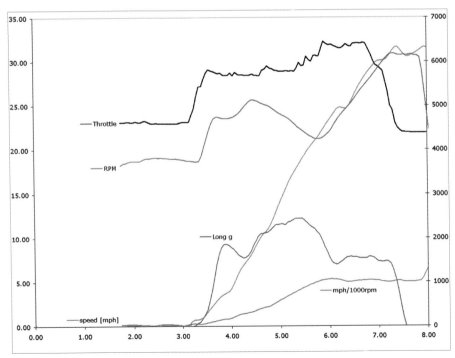

12.03. Using GPS and accelerometers gives different results.

existence to a wheel sensor that counts the pulses generated by the wheel studs passing a magnet. There is a significant jerk as the car leaves the line, and this is shown in the speed trace, which jumps to 6mph. The mph/1000 trace jumps vertically to start with, and then takes on the slope that indicates the driver is slipping the clutch. The long g figure is calculated from wheel speed, so it climbs rapidly to something over 4 g. If this is due to the driver's technique, it's pretty impressive, and something a dragster pilot wouldn't be ashamed of.

Sadly, though, it's a sensing issue. The speed sensor counts pulses, so when there are no pulses the trace shows zero. Now, if the car moves at a steady 1mph, that would be just over 0.4m/sec and with a tyre circumference of 1.5 metres, the wheel would be revolving at about one revolution every three seconds. A wheel stud would pass the magnet every 0.75 secs. At 2mph the studs would pass at about every 0.375sec, and so on. So, when the car moves, nothing much is going to happen until one stud passes the sensor. This will take a finite amount of time. The logger will then measure the time until the next stud passes, and calculate the speed that this indicates and begin to record a value for speed. The actual delay between launching the car and recording speed can be nearly half a second and any drag racer would regard that as an enormous delay. We haven't found a brilliant new launch method, only a snag with the way we measure it.

On the other hand, Chart 12.03 is compiled using a different system, with a GPS and physical accelerometers. In this case, the speed is shown in magenta, and mph/1000rpm in green. Both take off together as the car moves, and the initial upward slope of the green line indicates the extent to which the driver is blending clutch slip and wheel spin to make a good getaway. Toward the top of the chart, the red rpm and black throttle traces show that this was not simply a matter of dumping the clutch and allowing the wheelspin to do the job.

Dealing with heat presents the short-course racer with a slightly different set of problems to the circuit racer. On a longer event, the emphasis is on containing it within reasonable bounds; with events whose duration is measured in seconds or minutes, the problem is dealing with the adverse effects of not getting tyres and brakes up to temperature quickly enough, and judging the cooling requirements to a nicety so that you don't have to carry too much water or too big a set of heat exchangers. These are exactly the sort of questions that a logger can help you with. Brake and tyre temperature effects can be seen in the long g trace for getting off the line and braking for corners, and lateral g will tell us whether the tyres are giving more grip as they get hotter. At the other end of the run, you should be looking for deterioration due to heat effects and, if there is none, perhaps the signs are there that you could have gone softer or smaller with your engineering choices.

Unfortunately, measuring tyre and brake temperatures uses expensive technologies. Thermocouples for the brakes are pretty much standard industrial equipment, but their output is so low that the signal will need to be amplified to a suitable level. Tyre temperature can now be measured by infra red sensors, but they're not cheap, and will need a logger with a high channel count given the need for at least two extra sensors on each wheel.

The logger can also help with maximising the use of the tyres by using the traction circle as an analytical tool. Since the friction circle can be used to examine individual corners, it's possible to pay particular attention to whether there are benefits from braking right up to the apex of a corner, or maximising the traction under the exit. It's a revealing exercise, from the viewpoint of both chassis set-up and driving decisions.

There is a great benefit for drivers in being able to analyse in detail what went on in the short intensive experience that makes up a typical run. The blur of action can be augmented by a detailed examination of the data, and all the techniques suggested in Chapter 11 can contribute to the understanding that can help you to go faster.

KARTING

The similarities between karting and the circuit racing that forms the bulk of this book are obvious; the closed track, testing, qualifying and racing, and the need to optimise the whole package. The differences, too, are worth commenting on. The relative lack of power means that it's not only necessary to optimise the kart for the circuit, but to make sure to get the best out of the engine. Testing presents an opportunity to try different setups and the data should point you towards what works. The typical timetable of hourly practice sessions or races makes it essential to have a routine for getting information out of the logged data, so the first point of attention will depend on what the aim of the session was. If it was gearing adjustments, then rpm is the first thing to look at. If it was mixture, temperatures would be the focus, and so on. By the time of the race, the kart should be sorted and the emphasis should be on driving style.

Engine power

One of the major challenges of running a small engine is being sure that it's always getting the maximum horsepower from the engine, and choosing the right gearing to get the best from it. The data can help.

The best way to assess how close to optimal the air/fuel ratio is, is to measure either cylinder head or exhaust gas temperatures. You could log both, but just knowing one of them will help understand what's going on. Most direct is the exhaust gas temperature, because it's measuring the combustion process. Cylinder head temperature is influenced by other factors, such as airflow, so is only a surrogate measure.

As the mixture moves from rich to lean, the exhaust gas temperatures will first increase as the combustion becomes more complete, and then, once the ratio is reached for the fuel being used, they will start to fall away again. So, the peak gas temperatures will occur when the mixture is stoichiometrically correct – that is, when all the fuel is combined with all the free oxygen. This is great for fuel economy, but maximum power occurs with richer mixtures, so when peak temperature is reached, it's time to start going back down the scale towards somewhere where more power will be produced. What that means in terms of actual temperature depends on factors such as the ignition timing, exhaust scavenging, and compression ratio, so there's no magic number to aim for.

Cylinder head temperature works differently, and need treating with care as a tuning tool. Temperatures will rise with leaner mixtures, reach an optimum, and then you have to be careful because instead of falling they may get hotter still. What's happening here is that you may be seeing the effects of detonation – the combustion process getting out of control, and if you leave it like this you're likely to damage the engine. There's an element of feedback for the driver. With many karts it's possible to adjust the carb and change the mixture whilst on the move. Looking at the cylinder head temperatures gives the driver some idea of how effective the adjustments are.

Since there is a relationship between the exhaust gas temperature

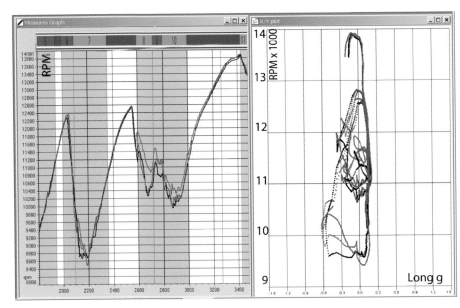

12.04. Assessing engine power with an XY graph.

and the correct air fuel mixture, and, therefore, the amount of power, it makes sense to use the logger to establish whether one setting is providing better power than another. Dynos are all very well, but they only measure the power output in the test cell, not on the track. Luckily, with short track karting (where aerodynamic forces are not so important) we can get a feel for how well the engine is performing.

The data that we need for this is engine rpm and long g. The trick is to focus on data from the straights and use it to create an XY chart of long g (acceleration) against rpm. In Chart 12.04 several laps are overlaid, and a part of the lap that includes major acceleration zones is selected. This section of the lap is plotted as an XY graph of long g (forward acceleration) against rpm.

The part of the XY graph that we are most interested in is the boundary towards the right of the chart which indicates the maximum acceleration. From this, it can be seen that the best acceleration is in the 11-12,500rpm range. There are two dips in the curve – in the 10–11,000 and 12,500 to

13,200rpm ranges. This gives us a very good idea of what we're working with. It's not a power curve, but it's a good indication of where the power is.

In testing, we could now change the jetting or exhaust length and run another few laps. Another XY graph, based on the same part of the lap will show whether we've shifted the acceleration boundary further to the right. It will also show in which direction the new settings move the boundary, and whether the way in which the power is delivered has moved up or down the rpm range, and whether there are any 'holes' in the shape of the curve. This technique will only work on short tracks with limited top speed.

The next stage would be to provide gearing that would keep the engine in the best part of the power curve. Success in this can be seen in an rpm histogram.

Gearing

The histogram is the tool for ensuring that the gearing is optimised. Chart 12.05 shows the bulk of the running in the 12,250 to 13,000rpm category, which fits quite well with what the XY chart

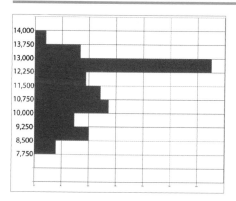

12.05. The histogram is vital in making gearing decisions.

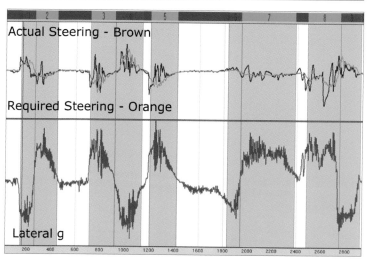

Actual Steering - Brown

Required Steering - Orange

Lateral g

12.06. The charts are the same as for a car, though the steering movements much exaggerated.

showed us. Here, there definitely seems to be a strong case for lowering the gearing, assuming that the engine will run happily to 14,000rpm, because we would get the benefit of a lower overall ratio and thus increase acceleration and also move the lower speed corners further up the power curve.

You need to think about not just a good spread of gearing over the whole lap, but whether you need to gear for one or more critical corners or straights. It's possible that you have a ratio that suits the whole track very well (evidenced by a good histogram and strong lap times) but that's hopeless for racing because it allows the opposition to overtake at the end of a long straight. Critical corners are often the long ones that the kart spends a lot of time in, or that lead on to an important straight.

Handling

All of the prescriptions for understanding the handling in the rest of the book apply to karting, except that the style of driving is more extreme here, and so the data can take on extreme forms as well. The niceties of interpretation are not available to you simply because of the karting driving style that requires the back axle to be made to slide to overcome the lack of a differential. Chart 12.06 shows a well behaved kart being driven with the necessary firmness. Turns 3 and 5

both seem to be taken with the steered wheels pointing in totally the wrong direction. The exaggerated steering movements associated with driving a kart can also be seen in Turns 4 and 8.

Slip

The fact that the kart has direct drive means that there has to be some slippage between the tyre and the road, and we can create a maths channel that will give us some insight into what is happening. To do this we need to compare speed with engine rpm and, in theory, should get a flat line because the two are tied together mechanically through the drive chain. If you gear to do 40mph at 10,000rpm then at 5000 you must be doing 20mph, and at 15,000, 60mph.

The reality is that the wheels spin (they need to in order to get round corners) and that the amount of spinning is consistent with getting good traction or braking. Tyres give their best grip when there is a small slippage between the road surface and the tyre contact patch. If this slippage gets too large, then the grip goes away. For example, under braking, a production car road tyre gives its best grip when the tyre speed is 20 to 30 per cent difference between the road speed and the tyre speed. There is a similar effect for acceleration. By calculating the slip, and sifting through the data, we can get a feel for what the optimal slip ratios are for a kart. This

could be done in a dimensionless form simply by creating a maths channel that is:

Slip = rpm ÷ speed

This would show slip as a number that would differ from kart to kart and according to the sprocket sizes. Its interpretation would be a matter of judgement. Much more satisfactory would be to calculate an actual ratio so that you could get some idea of the fit between your kart and the theory. This is slightly more complex because it involves knowing the actual gearing in use. Mph/1000rpm will tell us the actual gearing with the following formula:

mph/1000 = speed ÷(rpm ÷ 1000)

This can be plotted and a value read off the graph somewhere along a straight part of the circuit where there is no slippage. In Chart 12.07, this figure worked out at 3.78. The next stage would be to incorporate that number into another maths channel.

Slip = 3.78-(speed ÷ (engine÷1000)) *100

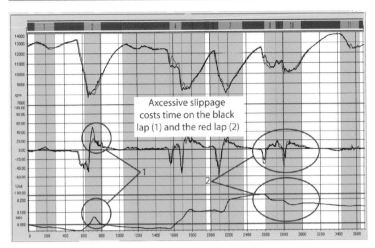

12.07. A chart like this helps determine how much wheel-slip is needed for fast times.

(chart label: Axcessive slippage costs time on the black lap (1) and the red lap (2))

This compares the nominal mph/1000 with the actual value calculated as the kart goes round the track, and multiplies it by 100 to provide a result as a percentage. Wheelspin under acceleration is a positive value, and wheels locking up under braking is negative. It's a clumsy way of doing things (having to enter a variable into a formula in this way), but it gives consistent results even where the gearing is different.

Chart 12.07 shows a plot in percentage terms that would help the driver to see the costs of excessive wheelspin, and the engineer to set up the kart to encourage or permit the correct level. Area 1 of the chart also shows the beneficial effects of slip because, although time was lost initially, there was a payback in terms of not letting the rpm fall, and keeping the engine in the power band that helped acceleration out of the turn.

RALLIES AND LOOSE SURFACES

Data logging for cars that run on loose surfaces has, until recently, been limited to engine logging. The engine health checks are vital and, in particular, teams keep a careful eye on the levels of boost, bearing in mind the arduous conditions under which the turbocharger is expected to operate. The reason for limiting logging to engine parameters has been the difficulty in measuring speed when the wheels are scrabbling for grip on loose surfaces. Without an accurate and clean speed trace, mapping is difficult, so understanding the path that the rally car takes is not possible. There have also been problems with needing timing beacons at the end of stages which might be some miles from the start. A racing team that thinks that placing a beacon on the start line is a difficult chore would give up completely with a rally special stage.

GPS has much to offer here. The maps are easy and accurate and need no external triggering. This makes it possible to use logging not merely to know what the engine is or has been doing, but now to look at the way in which the car is handling. Since few rallies take place on fresh roads all the time, you can learn lessons from the early stages for use later in the rally. Most events will re-use individual stages simply to avoid the effort involved in laying out fresh stages, and this helps with tyre choice. Logging adds an extra dimension to this because lateral acceleration data is indicative of the levels of grip being seen on the stages.

If the event consists of a series of short similar stages, the data gives a quick method of validating the decisions that have already been made, and adjusting the tactics if necessary. If the format is longer, then there's the opportunity for overnight analysis of the outcomes of the first day's sport.

Even though wheel speed sensors are likely to give odd readings because of the loose surfaces, there's a great deal of benefit that can be gained from recording all four speeds. They provide information about brake balance and the required balance ratio for various types of surfaces, and about traction. The extent to which there are differences in wheel speed can help understand the effectiveness of the differential(s) and the split between the front and rear. There's no need for complicated equations, simply subtracting left driving wheel from right driving wheel will evaluate the power distribution across an axle, and subtracting front speed from rear will tell us about the centre diff settings.

Tyre choice need not be simply a matter of matching the tyre to the surface, but if the team carries a range of tyres with different diameters it gives some flexibility in the choice of gearing. Competitors are normally given some idea of the length and shape of stages to come, so knowing that the next ones will be long and fast or short and twisty will give direction to the gearing.

Using the mph/1000rpm trace can be useful here as a quick indicator of tyre slip and, therefore, of the amount of grip that is available. Again, GPS systems are beneficial because, not only is the speed measurement more reliable than with wheel sensors, but also there is no need to re-calibrate the logger when the wheel or tyre sizes are changed.

Chapter 13
The field guide

The purpose of this chapter is to present a set of good and bad traces to give the reader chance to learn what shapes to look for and whether they are good news or bad. With practice, recognising the shape and its meaning will become as automatic as understanding the sounds that the engine makes or the way the car responds to the steering. Until then, here are

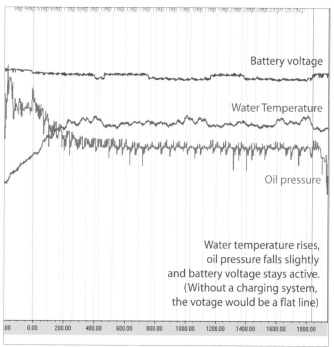

Battery voltage

Water Temperature

Oil pressure

Water temperature rises,
oil pressure falls slightly
and battery voltage stays active.
(Without a charging system,
the votage would be a flat line)

13.01. Good engine vital signs.

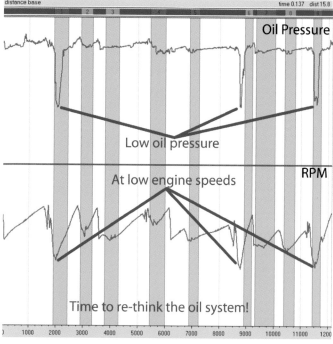

Oil Pressure

Low oil pressure

At low engine speeds

RPM

Time to re-think the oil system!

13.02. Low oil pressure.

some frequently seen shapes to practice on. Pack it with the check lists and the 'if this symptom-then take this action' list that should always accompany you to any event. It will speed up the problem solving process.

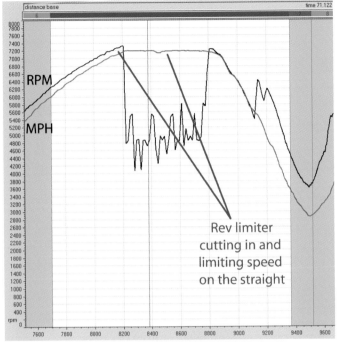

13.03. Rev limiter.

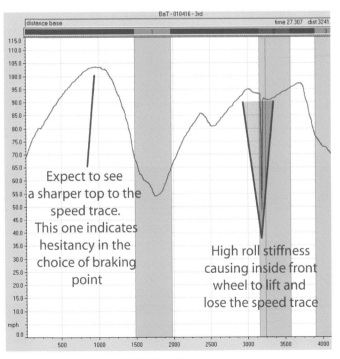

13.04. Good throttle.

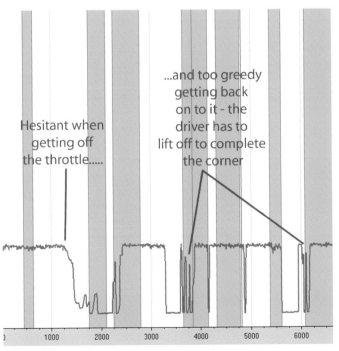

13.04a. Bad throttle.

13.05. Speed traces.

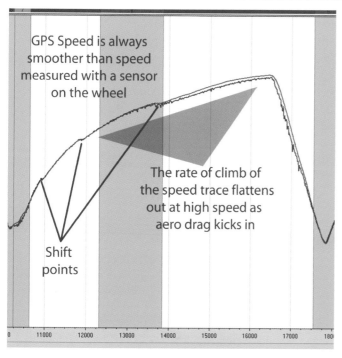

13.06. Speed.

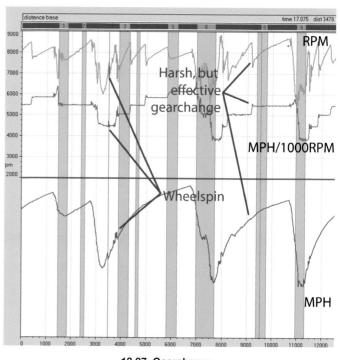

13.07. Gearchange.

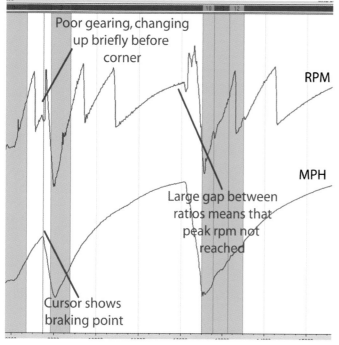

13.08. Gearing errors.

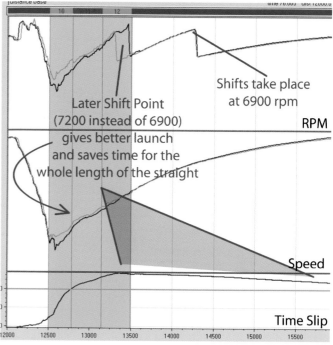

13.09. Shift point.

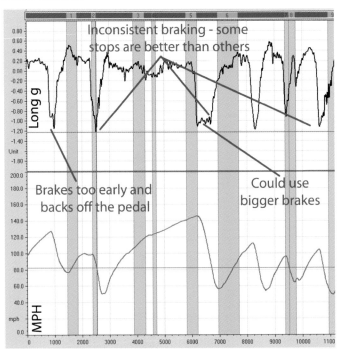

13.10. Braking.

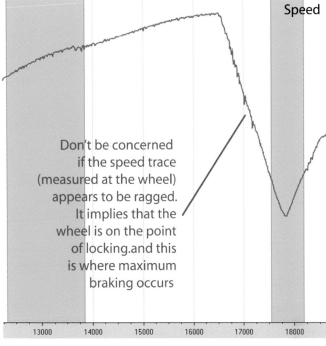

13.11. Braking well.

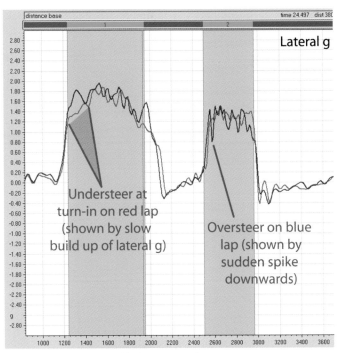

13.12. Understeer and oversteer.

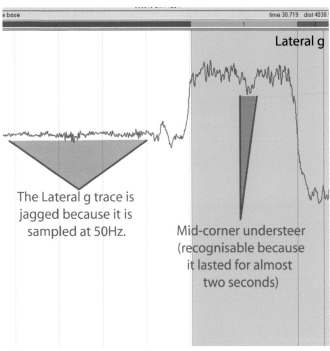

13.13. Understeer mid corner.

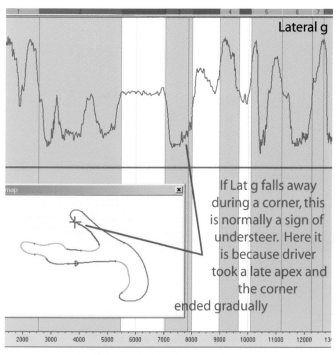

13.14. Lateral g that is not understeer.

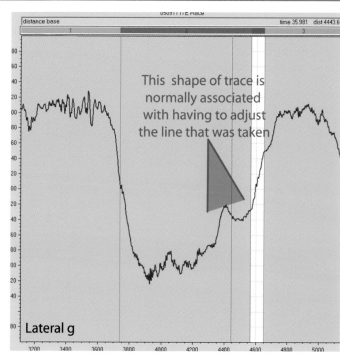

13.15. Adjustment to line.

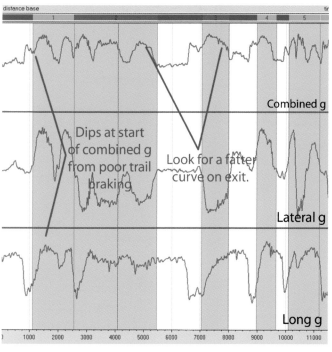

13.16. Combined g.

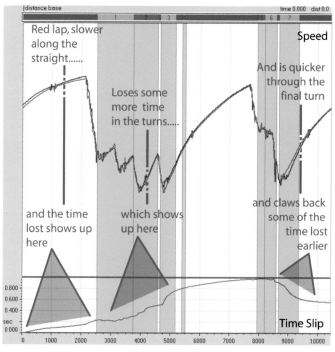

13.17. Time slip.

Appendix 1
Environmental protection ratings

This means that equipment with an IP65 rating would be totally protected against dust (first number – 6) and against low pressure low pressure jets from all directions with limited ingress of water (second number – 5)

	First number – protection against solids	Second number – protection against liquids
0	No protection	No protection
1	Protected against solid objects over 50mm	Protected against vertically falling drops of water
2	Protected against solid objects over 12mm	Protected against direct sprays up to 15 deg. from vertical
3	Protected against solid objects over 2.5mm	Protected against direct sprays up to 60 deg. from vertical
4	Protected against solid objects over 1mm	Protected against direct sprays from all directions - limited ingress permitted
5	Protected against dust – limited ingress permitted	Protected against low pressure jets from all directions - limited ingress permitted
6	Totally protected against dust	Protected against strong jets from all directions - limited ingress permitted
7		Protected against effects of immersion from 15cm – 1m
8		Protected against long periods of immersion under pressure

Appendix 2
Calibrating the sensors

Although most logger manufacturers provide methods for translating the sensor voltage into values in the data, there are still occasions when it's necessary to do this yourself. This Appendix shows you how to use Excel to derive functions that can be converted into maths expressions for use in the software.

CALIBRATING A SENSOR USING EXCEL

Once a piece of hardware has been installed, the software needs to be configured to translate its output into values that mean something to the team. In a perfect world, this would be a simple plug-and-play operation, and, in fact, this is one of the benefits that you get from buying an upmarket system. The software carries information about the range of common motorsport sensors, and you simply have to select the appropriate sensor from a drop down menu. One manufacturer even has a logger in the range that is marketed as

being suitable for newcomers to logging because it is completely pre-configured. This might be one of the things that influences your choice of system.

Most manufacturers provide some means of calibrating non-standard sensors by inputting the value in volts and the required data value. Going down the do-it-yourself route, and using something other than straightforward pressure, temperature or position sensors, means that there may be a few extra steps in getting useful information from the system. For instance, if we decide to set up a suspension position sensor based on a rotary pot, we need to convert the output from volts to millimetres of suspension travel. So, assuming that our sensor shows zero volts at full droop and 5V at fully compressed, our actions would depend on the software and how much work we want to do. Spending some time in the workshop would mean that we could provide a table like the one shown here.

It would be possible to simply do

Ride height	Voltage reading
-50	0
-40	0.70
-30	1.15
-20	1.52
0	2.15
20	2.60
30	2.93
40	3.22
50	3.62
60	4.12
70	4.76

nothing other than to plug the sensor into the appropriate channel of the logger and log suspension movement as 0 to 5. We would know then that ride height was about 2.5, and values greater than this meant that the suspension had compressed because of bumps or body roll. We could express body roll in volts

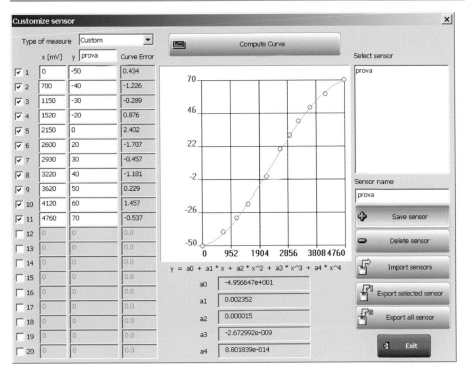

Screenshot 1. A manufacturer's calibration screen.

	A	B
1		
2	-50	0
3	-40	0.7
4	-30	1.15
5	-20	1.52
6	0	2.15
7	20	2.6
8	30	2.93
9	40	3.22
10	50	3.62
11	60	4.12
12	70	4.76
13		

Screenshot 2. The first step using Excel is to tabulate the data.

because we might only be interested in comparing data from one event to another.

This would be fine, because we could interpret what the wheels were doing, but if we had movement measured in millimetres, we could use that data to calculate body roll, get a rough estimate of downforce or aero lift, and a whole host of other interesting facts. As always, you get what you pay for, and the software from the more expensive systems will provide more help.

The easy way is to use a look-up table, although the manufacturer might not use this name. What you're looking for is somewhere in the software that allows you to enter readings (usually as millivolts so that 500mV = 0.5V) and measurements. This is then entered into the software and the logger can now provide data in millimetres of ride height. This is shown in screenshot 1.

If the software does not provide for direct entry of readings and values, it's still possible to get the measurements that you want if you can provide an equation that describes the relationship between input and output. One way to do this would be to dredge up high-school maths, but a much simpler method is to let a spreadsheet do the maths for you.

The starting point is to plot your measurements as an XY scatter chart. In Excel this involves creating a range of cells, as shown in screenshot 2, and going through the

Screenshot 3. Excel contains this Chart Wizard dialogue box.

step-by-step instructions to create a chart.

Using the menu options you select 'Insert Chart' and get the 'Chart Wizard' dialogue box. This is shown in screenshot 3. Here, an XY scatter has already been selected. The next step is to identify the data range, and a preview of the graph appears (this is shown in screenshot 4). There's no need to bother with the formatting options that you're

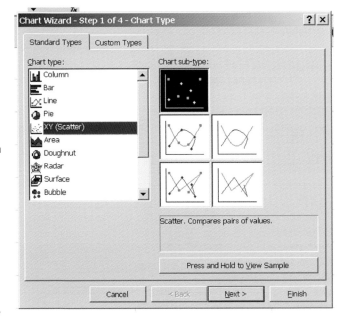

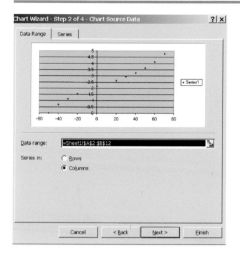

Screenshot 4. The preview of the graph.

presented with because the chart is just a part of the process of curve-fitting. A larger view of the graph can be seen in screenshot 5.

The clever bit is that Excel can insert a trend line. This means that it will look at the data that you supplied to create the graph, and work out the equation that best fits the shape of the curve. To do this, you need to select the curve by clicking on it with the mouse.

This operation is successful when the line changes to include the yellow blobs (screenshot 6). Right clicking at this point brings up a sub-menu which offers you the chance to add a trend line. Select this and you're presented with yet another window that gives you several alternative types of trend line. If your data is a straight line (as should be the case with steering angles) select linear. If the line is curved, generally speaking a polynomial expression will define the curve best (this is shown in screenshot 7, but there has been a bit of cheating here and, after a bit of trial and error, the third order polynomial was found to fit the curve best). In most logging instances, second or third order polynomials will suffice. Select the options tab and choose to display the equation on the chart; the end product should look something like screenshot 8.

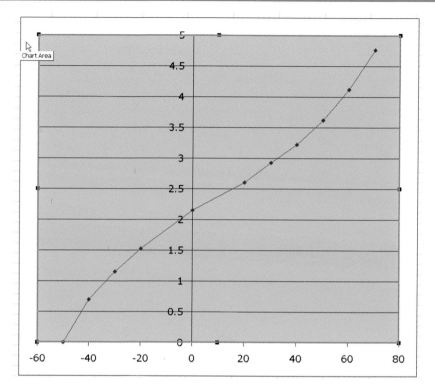

Screenshot 5. A larger view of the graph.

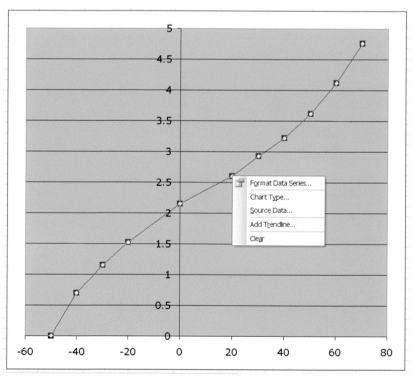

Screenshot 6. Selecting the line to create a trend line.

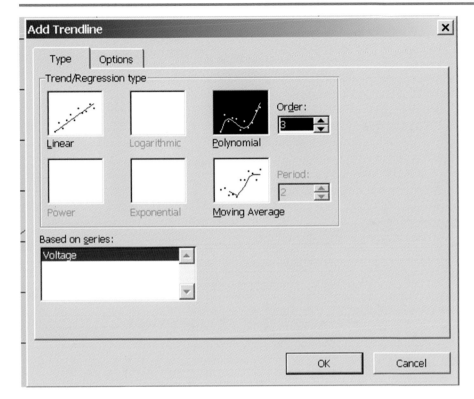

Although it is not exactly presentation-standard graphics, it is the equation that we are after. The x is the voltage and the y is the measurement. So, for the datalogging purposes, the x is the value of the variable recorded by the logger, and the y is the number that we want to display in our charts and calculations. The numbers in this formula differ from those calculated in the AIM software in screenshot 1 only because the AIM uses a fourth-order polynomial expression, whereas in Excel we used a third-order polynomial. The difference is too small to be significant.

The equation for the trend line is:

$$y = 4E\text{-}06x^3 - 0.0001x^2 + 0.0253x + 2.1389$$

and this looks pretty complicated. The first thing that you notice is that it uses the exponential notation so that 4E-6 means that the 4 is the sixth digit in the number and the - indicates that there are leading zeroes. 4E-6 means the same as 0.000004.

The $4E\text{-}6x^3$ part of the equation means that you take the voltage value (x) raise it to the power of three and multiply by 0.000004. The equation is completed by taking away the x^2 multiplied by 0.0001, adding x times 0.0253 and finally adding 2.1389.

In the computer notation that your logging software will understand, this would be expressed as:

0.000004*(Channel_name^3) – (0.0001* Channel_name^2) + (0.0253*Channel_name) + 2.1389.

There are several quirks in this formula that, perhaps, deserve some further explanation. The word

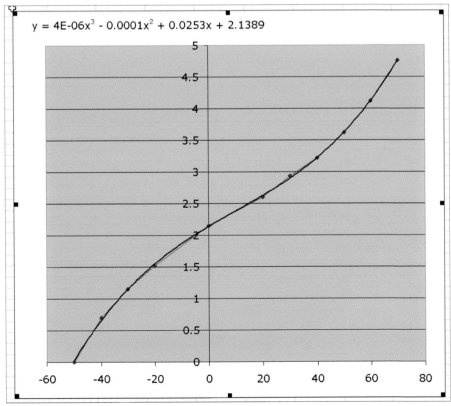

$$y = 4E\text{-}06x^3 - 0.0001x^2 + 0.0253x + 2.1389$$

Channel_name is used to represent the variable name that the logger uses; the protocol varies from system to system, so you'll need to identify the name that your system is using for this sensor. The ^ symbol indicates raising to the power so 2^4 is computer speak for two to the power 4.

Brackets are used to make sure that the computer does the calculations in the correct order. For example, the answer to 2 + 3 x 4 can be either 20 or 14 depending on whether you do 2+3 = 5 x 4 = 20. But if you do 2 plus the result from 3 x 4 (12), the answer is 14. The normal order for a computer to do its calculations is numbers in brackets first, then raising to a power, multiplication, division, addition and subtraction.

When Excel inserts a trend line it tends to do it only to 2 decimal places. If the behaviour is at all complex this will need to be expanded to more decimal places. You can do this by right clicking on the value in the chart and formatting the number to as many decimal places as necessary.

So, if the software does not provide the facilities for custom calibration of sensors, the method outlined here provides a way of doing so. The stages are first to create a table showing the real-world units (millimetres, degrees, etc.) and the logger voltages that represent them. Use Excel to draw a graph of your findings and to fit a trend line. Use the equation that represents this trend line to create a maths channel where y is the value that you want to display in the logging software and x is the value that the logger measures.

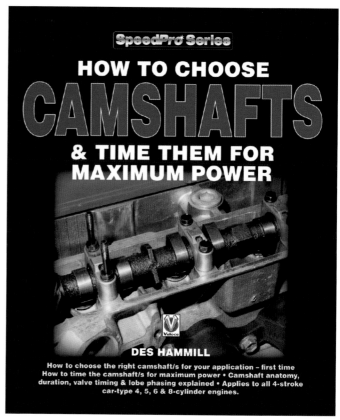
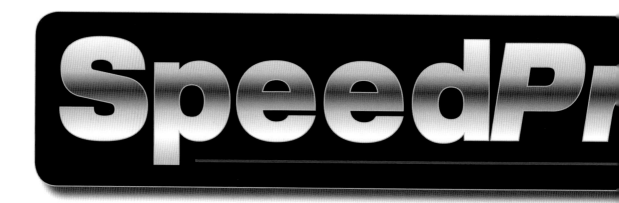

SpeedPro Series

THE **SPORTS CAR** & **KIT CAR**
SUSPENSION
& BRAKES
HIGH-PERFORMANCE MANUAL

DES HAMMILL
How to build & modify • Applies to all sports cars & kit cars with wishbone front suspension, coil springs, telescopic shock absorbers & wishbone or trailing arm rear suspension.

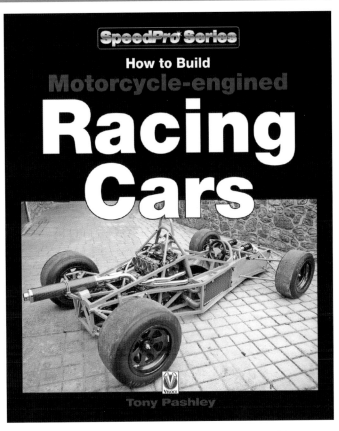

SpeedPro Series

How to Build
Motorcycle-engined
Racing Cars

Tony Pashley

ISBN: 978-1-845842-07-9 • £19.99*   ISBN: 978-1-845841-23-2 • £24.99*

Note: all prices subject to change • P&P extra. Call +44 (0)1305 260068, or email sales@ veloce.co.uk for more information.

www.veloce.co.uk

ALSO FROM VELOCE PUBLISHING –

ISBN: 978-1-845841-86-7 • £16.99*

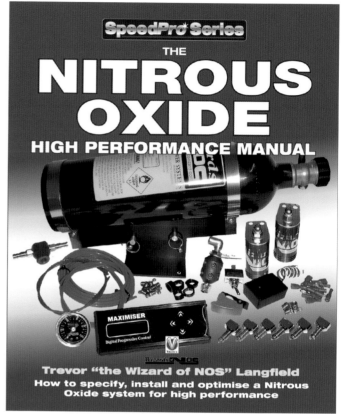

ISBN: 978-1-904788-89-8 • £15.99*

Note: all prices subject to change • P&P extra. Call +44 (0)1305 260068, or email sales@ veloce.co.uk for more information.

www.veloce.co.uk

ISBN: 978-1-845841-03-4 • £17.99*

ISBN: 978-1-845841-66-9 • £19.99*

Note: all prices subject to change • P&P extra. Call +44 (0)1305 260068, or email sales@ veloce.co.uk for more information.

www.veloce.co.uk

Index